SUPERCONDUCTIVITY BEGINS WITH *H*

both properly understood, and misunderstood

Superconductivity basics rethought

SUPERCONDUCTIVITY BEGINS WITH *H*

both properly understood, and misunderstood
Superconductivity basics rethought

Jorge E. Hirsch
University of California San Diego, USA

World Scientific

NEW JERSEY • LONDON • SINGAPORE • BEIJING • SHANGHAI • HONG KONG • TAIPEI • CHENNAI • TOKYO

Published by

World Scientific Publishing Co. Pte. Ltd.
5 Toh Tuck Link, Singapore 596224
USA office: 27 Warren Street, Suite 401-402, Hackensack, NJ 07601
UK office: 57 Shelton Street, Covent Garden, London WC2H 9HE

British Library Cataloguing-in-Publication Data
A catalogue record for this book is available from the British Library.

First published 2020 (hardcover)
Reprinted 2026 (in paperback edition)
ISBN 978-981-98-3788-5 (pbk)

SUPERCONDUCTIVITY BEGINS WITH *H*
both properly understood, and misunderstood: Superconductivity basics rethought

ISBN 978-981-121-685-5 (hardcover)
ISBN 978-981-121-686-2 (ebook for institutions)
ISBN 978-981-121-687-9 (ebook for individuals)

For any available supplementary material, please visit
https://www.worldscientific.com/worldscibooks/10.1142/11734#t=suppl

Desk Editor: Nur Syarfeena Binte Mohd Fauzi

Typeset by Stallion Press
Email: enquiries@stallionpress.com

Preface

Superconductivity is one of the most interesting phenomena in physics. It is also one of the most controversial fields of physics, but even so, I believe many would agree with the first statement. Let me explain why.

Superconductivity is a *bridge* between macroscopic physics, with which we are familiar in our everyday life (Newton's laws of mechanics, thermodynamics, Maxwell's electromagnetism), and quantum mechanics, whose laws govern the microscopic behavior of atoms and molecules, which is much less intuitive and familiar to (almost all of) us. Superconductors have been characterized as 'giant atoms'. The electron in the microscopic hydrogen atom orbits around the nucleus indefinitely without losing energy due to friction. Such is also the behavior of the electric current in a superconducting ring of, for example, 1 cm diameter. The current has been measured and it persists for years without decaying, one expects that it will exist forever. Very different from any other motion at the macroscopic scale that experiences friction and decays with time unless energy is continuously supplied to keep it going. This attribute of superconductivity of being a 'bridge' between microscopic and macroscopic realms is unique and makes the topic particularly fascinating.

It also has important consequences due to what is known as 'Bohr's correspondence principle'. In 1913, Niels Bohr relied on the principle that microscopic physics should smoothly evolve into macroscopic physics as the dimensions of the system grow in order to deduce the laws that govern the behavior of the hydrogen atom. Surprisingly, however, this principle is not taken into account in the contemporary science of superconductivity. It relies on microscopic physics and disregards Bohr's correspondence principle. This is one of the main reasons why I am convinced that superconductivity as it is understood today is misunderstood.

If superconductors could be used in everyday life, energy and pollution problems would be notably reduced, dependence on oil would be reduced, and problems in the Middle East would be solved. It is an important subject!

In this book, as I have for many years in scientific articles, I propose that the main player in a superconductor is the *hole*. Holes are positive charge carriers or electricity rather than negative as the electron is. The concept of holes in atoms, molecules, and solids was introduced by Werner Heisenberg, one of the founders of quantum mechanics, in 1931, and it means the absence of an electron. In the accepted theory of superconductivity, holes don't play any special role. Instead, for the last 30 years, I have advocated that holes play a fundamental role in superconductivity and that without holes, there cannot be superconductivity. That is the main reason for the title of this book: *Superconductivity (properly understood) begins with H*, *H*eisenberg's *H*oles.

Why do I say in the title that superconductivity *misunderstood* also begins with *H*? It is in reference to *H*erbert Fröhlich, a German physicist who in the year 1950 published an article that conditioned the development of the theory and understanding of superconductivity ever since. In that article, Fröhlich proposed that what causes superconductivity is the interaction between electrons and the vibration of the atoms in the solid. The entire evolution of the theory, until today, is based on that principle, if that principle is wrong, the theory collapses. I say not only that Herbert Fröhlich proposed a wrong principle, but also that he did it in a scientifically *dishonest* way, as I will explain in this book. This fact crucially determined the evolution of this field of science in the wrong direction, leading to the misunderstanding of superconductivity that exists today. For this reason, *misunderstood superconductivity begins with H*erbert Fröhlich.

There is a second reason for why I say that superconductivity *misunderstood* begins with *H*. The *H index* is a bibliometric index that I proposed in 2005 to measure the scientific quality of researchers based on the citations their papers receive. It is generally accepted as a good indicator: highly recognized scientists have high *H* indices and vice versa. In solid state physics, many of the most prominent scientists working on superconductivity based on the accepted theory of superconductivity have very high *H* indices. Against my own proposal of 2005, I have to say that in those cases, a high *H* index does not reflect scientific quality, and it discourages young scientists to raise questions about the established theory because they assume that if high *H* index scientists promote it, it has to be correct. Therefore, *Superconductivity misunderstood* (and mistaught and that way propagated) *begins with* scientists of high *H* index.

Returning to superconductivity *properly understood*, here is another reason for the title of this book. To properly understand superconductivity, we have to start questioning the understanding proposed by the established theory of superconductivity, theory that is considered by the scientific community to be an absolute truth, undeniable, and unquestionable, like a religion. *Superconductivity (properly understood) begins with H*eresy.

Briefly about myself: I graduated with a 'Licenciado en fisica' ('licenciate in physics') degree in 1974 at the University of Buenos Aires, and obtained my Ph.D. in physics at the University of Chicago in 1980. Since 1983 to the present, I have taught and performed research at the University of California, San Diego. I have always worked in solid state physics and published over 200 articles in scientific journals. Some were highly read and cited, others (the best among them in my opinion) not. In addition, I have done some work on bibliometrics, I published four articles on that subject, which is 1.6% of my scientific production that have in total 22% of the total citations to my work.

Despite having done physics practically all my life in English, I first wrote this book in Spanish because it is my native tongue and I found it easier to express in that language issues about which I feel strongly. The title (In Spanish "*La superconductividad bien entendida empieza con H*") also expressed my belief that, during the time that the book would remained untranslated, *Superconductivity properly understood begins with H*ispanic readers. After reading this book, readers will understand fundamental things about superconductivity that 99.99% of the scientists expert in superconductivity don't. In particular, how the Meissner effect works, the most fundamental property of superconductors.

The potential readers I mostly had in mind while I wrote this book know some science. They learned, either in high school or in undergraduate studies in engineering, natural sciences or medicine, Newton's equations of mechanics, Maxwell's laws of electricity and magnetism, as well as the first and second laws of thermodynamics. Also, they know at least qualitatively the principles of quantum mechanics, if only because of having read a popular science book on the subject. My objective is that those readers if necessary refresh their memory of these physics principles and understand all physical arguments and mathematical formulas presented in this book. With those elements, they can make an informed judgement on whether what I say is possible and likely to be true, or not. Reading this book, I hope they will have an enjoyable time and learn things that will be of use to them in the future.

On both sides of that ideal reader, I think there are also potential readers that can benefit from reading this book. Those that know nothing about physics will have to skip the mathematical formulas, but even so I think they can understand qualitatively the majority of arguments that I present, in fact the most important ones are so simple that they can be understood without any scientific background. On the other hand, there will be those readers who know a lot of physics, maybe even some that are experts in superconductivity. I hope those readers will read this book with their minds open to the possibility that much of what they learned in other books and articles on superconductivity is possibly not right, and will let me know whether or not this book changed their views on the subject.

In addition, I think all readers can find of interest the sociological aspects of the subject I develop, related to how science advances and stalls and how the tendency to preserve the *status quo* can seriously hamper scientific development.

More than anything, I hope that this book will stimulate some readers to contribute to the understanding of superconductivity, in that way accelerating the advent of a time in the future when this singular and fascinating phenomenon will occupy the prominent place in science and technology that it deserves, has promised for a long time, but has as yet not reached.

Contents

Part I

OVERVIEW OF SUPERCONDUCTIVITY

Chapter 1

Introduction

As I mentioned in the Preface, superconductors behave in a way like 'giant atoms': electrons circulate in persistent currents without dissipating energy, as shown schematically in Fig. 1.1.

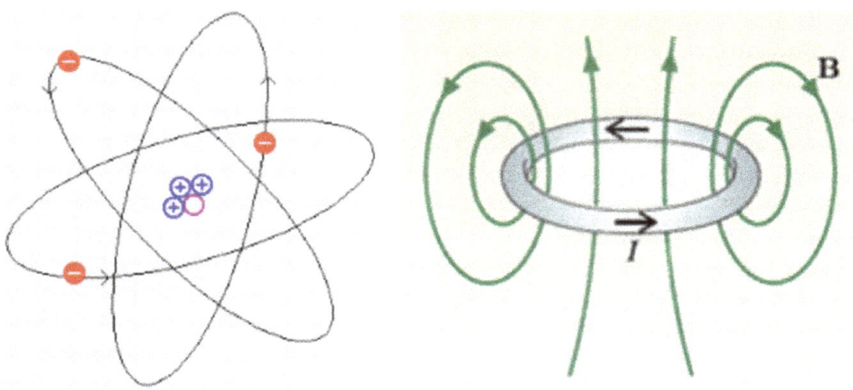

Fig. 1.1 Left panel: electrons in the atom orbit continuously without dissipating energy. Right panel: in a superconducting ring with current I, electrons move continuously without dissipating energy. The lines of B denote the magnetic field generated by the current. The scale is 1 cm on the right, 0.000000001 cm on the left.

Superconducting materials have two fundamental properties when they are cooled below their 'critical temperature' T_c that varies with each material: (1) The electrical resistance suddenly vanishes, that is, electric current flows without resistance and without any energy loss. This was discovered in 1911 by Kammerlingh Onnes. (2) If the material is located close to a magnet, in the normal state (i.e. at temperature $T > T_c$), there is essentially no force between the material and the magnet (we assume the material is not a ferromagnetic metal like iron or nickel, in that case, there would be a force); when the material is cooled below its critical temperature, a repulsive force between the material that became a superconductor and the magnet suddenly appears. This is because an electric current starts flowing in the superconductor that generates a magnetic field opposed to the magnetic field of the magnet. This is the Meissner effect, discovered by Meissner and Ochsenfeld in 1933, that is 22 years after the discovery of the property (1). We will return to discuss the Meissner effect many times in this book due to its fundamental importance.

In the year 1957, a theory was formulated to explain superconductivity, BCS theory, that supposedly explains the phenomena described above. This theory, due to physicists Bardeen, Cooper, and Schrieffer (hence BCS) (awarded the 1972 Physics Nobel Prize for this achievement), also known as the 'conventional theory of superconductivity', is universally accepted as valid to describe the majority of superconducting materials, the so-called 'conventional superconductors'. They include all the superconductors known in 1957, in particular, the simplest materials, superconducting elements (e.g. lead, niobium, aluminum, tin), and simple compounds and alloys of two or three elements. There are other superconducting materials discovered in the last 40 years known as 'unconventional superconductors', for example, copper oxides, for which it is generally agreed that they are not described by BCS theory, but there is no agreement on an alternative theory to describe them. In general, these 'unconventional' materials superconduct at temperatures higher than the conventional ones, hence are more interesting from a technological point of view.

The 'Holy Grail' of superconductivity is to find materials that superconduct at room temperature and atmospheric pressure, which would enormously broaden the use of superconductors for practical applications, with great benefit to society. Currently, the 'hottest' superconductor at atmospheric pressure, $HgBa_2Ca_2Cu_3O_{8+x}$ (also known as Hg1223), superconducts at ~135 K, i.e. −138°C, i.e. 138°C below the freezing temperature of water. There is some distance to be traveled!

In this book, I will explain why I am convinced that the universally accepted theory of conventional superconductors, BCS, is wrong. It is not that it is all wrong, but certainly sufficiently wrong to be incapable of predicting which materials are superconductors and which are not. This is a serious defect, since it doesn't help us in the search for higher temperature superconductors. More importantly, BCS is wrong because it doesn't explain the Meissner effect described above, one of the two fundamental properties of superconductors. The fact that BCS is wrong, but everybody believes it is right, is hampering the advancement of the science of superconductivity and its technological applications to the detriment of science and society. But since the theory is universally believed to be correct, to say that it is wrong is considered heresy in the scientific world.

For the so-called 'unconventional superconductors', there is no universally accepted theory. But a great majority of scientists focus their attention on a model describing interacting electrons in solids known as the 'Hubbard model', which is considered to have the essential elements to explain 'unconventional superconductivity'. I myself did research on the Hubbard model many years ago, in fact I was one of the first (if not the first) to propose that it explains 'unconventional superconductivity', in articles I wrote in the 1980s that are still highly cited today. However, shortly thereafter, I came to the conclusion that the Hubbard model does not explain superconductivity in any material and wrote several articles explaining why. Nevertheless, the focus of the scientific community studying superconductivity in the so-called 'unconventional' materials continues to be the Hubbard model. Another reason for the title of this book: *Misunderstood* 'unconventional' *superconductivity begins with* the *H*ubbard model.

For the last 30 years, I have been convinced that there is a single explanation of superconductivity, which applies equally to the so-called 'conventional' and 'unconventional' superconductors. Obviously, if this is correct, it implies that BCS theory cannot be correct. This generates extreme opposition in the scientific community, which considers BCS theory to be the greatest scientific achievement in solid state physics in the last 80 years. The main reason why I am convinced of this is because abandoning the devotion to BCS, one can explain the Meissner effect, as well as many other properties of superconductors, in a simple and intuitive way, as I will explain in this book.

I call the alternative theory that I and coworkers have been developing for over 30 years: '*hole superconductivity*'. It has essential physical elements that are not part of BCS theory and some that are. In addition to applying to all materials, another advantage over BCS is that it is simpler and more intuitive, easier to understand and explain, as I will attempt to do in this book. This theory is not considered valid not for any concrete scientific reason, but rather because it questions the validity of BCS. This is not acceptable in contemporary physics, just as it was not possible in the middle ages to question the absolute validity of the church's precepts.

Fortunately, we don't live in the middle ages. The reader can decide.

Chapter 2

Bird's eye view of superconductivity: Heroes and villains

As mentioned in Chapter 1, Kammerlingh Onnes, a Dutch physicist, discovered superconductivity in 1911, 3 years after having achieved the liquefaction of helium, which allowed him to reach temperatures lower than those ever reached before. Onnes wanted to know what happens to the resistance of metals when they are cooled to close to absolute zero. There were two competing hypotheses: (1) electrons would 'freeze' and the resistivity would become infinite or (2) the ionic motion would gradually freeze, and since thermal vibration of ions causes resistance as electrons collide with ions, the resistance would gradually approach zero. Surprisingly, Onnes found something nobody expected: at a critical temperature of $4.5\,\mathrm{K}$ (4.5° above absolute zero), the resistance of mercury dropped from a finite value to zero.

Onnes immediately recognized the fundamental importance of his discovery, and he continued studying the phenomenon intensely. Shortly thereafter, he discovered several other metals, as well as binary compounds, that exhibited the same behavior, each at its own critical temperature. Doing experiments with magnetic fields, in 1914 he established the existence of a critical magnetic field, distinct for each material, above which the superconducting state is destroyed. He also demonstrated the existence of a critical maximum current that a superconductor can carry that depends on the material. All were very important discoveries.

Fig. 2.1 Heroes of superconductivity: from left to right, H. Kammerlingh Onnes, Walther Meissner, Fritz London, Heinz London, and John Bardeen.

But Onnes also made a mistake. In 1924, he found that in cooling a metal sphere in the presence of a magnetic field, when it went from the normal to the superconducting state the magnetic field did not change. It didn't surprise him, it was what was expected according to the theory of electromagnetism. But the result was wrong, and Onnes missed the chance to discover another fundamental effect, the Meissner effect.

Of course, Onnes was indisputably a hero. *Superconductivity properly understood begins with H*eike Kammerlingh Onnes, certainly. More than deservingly, he was awarded the 1913 physics Nobel prize, for superconductivity and other scientific achievements.

The next experimental step of fundamental importance was the 1933 discovery by Walther Meissner and Robert Ochsenfeld, German physicists, that if a metal becomes superconducting in the presence of a magnetic field, an electric current spontaneously starts flowing near the surface of the metal, which creates a magnetic field in the interior of the material that exactly cancels the magnetic field that existed previously, resulting in zero magnetic field in the interior of the superconductor. In other words, the magnetic field is expelled from the interior of the superconductor. This causes the repulsive force between the superconductor and the magnet mentioned in Chapter 1. The discovery of the Meissner effect was entirely unexpected, since it was expected that magnetic fields cannot be modified inside a perfect conductor (i.e. a conductor with zero resistance) according to Maxwell's equations of electromagnetism.

Why had Kammerlingh Onnes missed this effect in his 1924 experiment? First of all, he wasn't looking for it, nobody expected it.

Second, to save on the amount of liquid helium necessary to cool the superconductor, Onnes had used a hollow rather than a solid sphere. For a hollow sphere with thin walls, the Meissner effect is very small, and Onnes didn't detect it.

Walther Meissner too was definitely a hero. Deserving of a Nobel Prize, he was nominated 6 times but never received it, despite having lived till 1974, 2 years after Bardeen–Cooper–Schrieffer (BCS) were awarded their Nobel Prize. An injustice.

Immediately after 1933, the London brothers Fritz and Heinz, German Jews exiled in England due to Hitler's rise to power, formulated a phenomenological theory of superconductivity that describes the Meissner effect and in addition suggests that superconductivity is a macroscopic quantum effect, with a well-defined *quantum phase*. This work, based on the two 'London equations' that they introduced, was the fundamental basis of all the subsequent theoretical developments, including BCS. Fritz London wrote a book in 1950, *Macroscopic Theory of Superconductivity*, where he masterfully reviewed what was understood about superconductivity at the time, mostly due to him and his brother. There he also predicted the quantization of magnetic flux in superconducting rings, a fundamental phenomenon verified experimentally only many years later.

Fritz and Heinz London were definitely heroes too, especially Fritz. With a caveat however: I argue that half of London's equations, i.e. one of their two equations, is not correct. On the other hand, this is amply compensated by another contribution of the London brothers that is largely unknown, which they themselves thought was wrong but which in fact is right and is important. It is not described in books on superconductivity, not even in London's 1950 book. I will explain this later.

There were a multitude of theoretical physicists who tried to explain superconductivity between the years 1911 and 1957. I believe several of them had correct and important ideas, which were largely ignored at the time they were proposed and definitively discarded when the BCS theory was introduced. I mention in particular physicists Frenkel, Landau, Kronig, Born and Cheng, Heisenberg, Smith and Wilhelm, Koch, and Slater. They were also heroes, of lesser importance than the ones mentioned earlier, who did not receive the recognition they deserved. Einstein himself also tried to contribute to the understanding of superconductivity with a small paper in 1921, which was deservedly ignored.

Finally, John Bardeen, an American physicist, the first author in BCS, is also one of the major heroes. He devoted himself intensely to understanding superconductivity, starting in 1950, shortly after having made fundamental contributions to the science of semiconducting materials that earned him his first physics Nobel Prize in 1956 (the second one was for BCS in 1972). In 1956, Bardeen wrote a review article on superconductivity defining clearly the problems that in his view needed to be solved. Shortly thereafter, together with his collaborators Leon Cooper (post-doctoral researcher) and Bob Schrieffer (graduate student), he formulated in 1957 the BCS theory, which has many novel ingredients that explain many aspects of superconductivity. Unfortunately, I believe it also has many ingredients that are incorrect, even though they are considered correct by everybody else, as well as many missing crucial ingredients that are necessary to describe and explain superconductivity. In any event, despite these caveats, John Bardeen is definitely a hero in this story.

In this way, we complete the list of heroes of the early developments in the understanding of superconductivity. We can say without doubt that *superconductivity properly understood begins with* these *H*eroes.

The first villain (and a major one) in the history of superconductivity was Herbert Fröhlich, mentioned in the Preface, another German Jew exiled in England due to Hitler. In 1950, Fröhlich proposed that superconductivity originates in the interaction between electrons and lattice vibrations of the ions (called phonons). This was in appearance confirmed by the simultaneous discovery of the isotope effect, the variation of the critical temperature of superconductors with different isotopic mass of the ions, by E. Maxwell, B. Serin, and collaborators. Since then up to the present, it has been universally believed that this interaction, known as 'electron–phonon interaction', causes superconductivity in 'conventional superconductors' (not in the 'unconventional' ones). I believe that this is incorrect, that the electron–phonon interaction does not cause superconductivity in any material. Complete heresy for BCS devotees.

Herbert Fröhlich was definitely a villain. Later I will explain in detail why.

Other villains in the history of superconductivity? All that follow BCS theory and consider it unquestionable, despite all the evidence against it that has been accumulating over the years. The greatest villains are the most prominent scientists in this group who refuse to consider the possibility that BCS is not completely correct, they suppress any way they can the expression of ideas that may question BCS and teach their many disciples to do the same. Because of this, papers questioning BCS cannot be published in the most prestigious scientific journals, nor can such ideas be presented at the major scientific meetings.

The list of villains is very long. I mention here only some of the more prominent ones with whom I have interacted (Fig. 2.2): Neil Ashcroft, Doug Scalapino, Phil Anderson, and Marvin Cohen. They are all very prominent scientists in the United States, members of the National Academy of Sciences, and with very high H-index: *60, 97, 115, and 122*, respectively. Much higher (except Ashcroft's) than mine, *58*. The H-index measures how many papers of an author have more citations than that number. For example, 97 papers of Scalapino have each more than 97 citations. That represents an extraordinary scientific production, highly recognized. But, a large part of that production is on the subject of superconductivity, that

Fig. 2.2 Some villains of the theory of superconductivity: from left to right, Herbert Fröhlich, Neil Ashcroft, Doug Scalapino, Phil Anderson, and Marvin Cohen.

I say is misunderstood by these prominent scientists. *Superconductivity misunderstood begins with* scientists of very high H-index.

More specifically, who are these 'villains'?

Neil Ashcroft proposed in 1968, based on BCS, that the simplest element, hydrogen, subject to very high pressure, would become a superconductor at high temperatures. He continued insisting on this throughout the years, and has many followers. In recent years, several articles have been published claiming that they detected high temperature superconductivity in hydrogen-rich compounds at very high pressures. I say that this is impossible, since one would not expect to have 'holes' in hydrogen. For that reason, *superconductivity misunderstood* also *begins with H*ydrogen.

Phil Anderson worked on BCS theory shortly after it was published in 1957, and found arguments to support it. Much later, in 1987, shortly after the discovery of high temperature cuprate superconductors, he published a very influential article proposing that the explanation of the superconductivity of the cuprates is that they form what is known as a 'spin liquid', described by the Hubbard model. This caused a large part of the scientific community to concentrate on studying spin liquids and the Hubbard model in the ensuing 30 years, until the present. I say that this has nothing to do with the superconductivity of the cuprates, that *Superconductivity misunderstood begins with* the Hubbard model. Why has Anderson's proposal had such a big influence? In part surely because he is the most prestigious condensed matter physicist alive today, recipient of the 1977 physics Nobel prize. As he immodestly states in his webpage, *I am a condensed matter theorist, a field in which I played the role of a major agenda-setter for 40 or so years.*

Doug Scalapino disagrees with the spin liquid concept, but he also believes that the Hubbard model explains the cuprates for other reasons, and has devoted enormous efforts during the last 30 years with his many students and collaborators in trying to prove this, by performing numerical simulations on the most powerful supercomputers, publishing an enormous number of articles with results that in my opinion are questionable and/or irrelevant to the physics of superconductivity in the cuprates. Again, this is *Superconductivity misunderstood*. Scalapino also made early contributions to BCS theory in the 1960s that are highly recognized by the scientific community despite contributing, in my opinion, to the misunderstanding of superconductivity, as I will discuss later.

Marvin Cohen, since the 1960s, has been performing 'first-principles' calculations where he purports to calculate the critical temperature of conventional superconductors without using experimental data. The problem is, he always obtains the result after the critical temperature has been measured, he never predicts a critical temperature before it is measured. Notwithstanding his claims to the contrary, e.g. in an article he published a few years ago "Predicting and explaining T_c and other properties of BCS superconductors", where he discusses his work on this over many years. His papers are highly cited and recognized.

More about these and other villains in subsequent chapters.

There is one important hero that I have not yet mentioned, from the far past: Bernd Matthias.

Fig. 2.3 Bernd Matthias, hero of superconducting materials.

Matthias, another German Jew exiled first to Switzerland and then to America in 1947, was an experimental physicist who started to work in superconductivity in the early 1950s, before BCS, and continued intensely until 1980, when he died suddenly from a heart attack. Coincidentally, from 1963 to 1980 he was professor at the same university where I have worked from 1983 to the present, University of California San Diego (UCSD). Matthias discovered many superconducting materials using empirical criteria that he developed, and was always very skeptical of BCS theory. He was the only scientist to publicly question the validity of BCS theory in the period 1957–1980. Matthias did think that BCS applied to some conventional materials, but thought there were other mechanisms besides electron–phonon that were the cause of superconductivity in many other materials considered to be conventional, including some elements. Matthias died just around the time that materials now considered to be unconventional, i.e. believed not to be described by BCS and where superconductivity is not due to the electron–phonon interaction, started to be discovered.

Bernd Matthias (Fig. 2.3), besides being a *h*ero, was also a *h*eretic who contributed to *superconductivity properly understood.* Since Matthias's passing, nobody else has questioned the validity of BCS theory for conventional superconductors. Even today, nobody other than myself questions it.

From the recent past, we need to highlight the heroes of high-temperature superconductivity (Fig. 2.4): Alex Müller and Georg Bednorz, Swiss, and Paul Chu, Chinese-American.

Fig. 2.4 Alex Müller, Georg Bednorz, and Paul Chu, heroes of high temperature superconductivity.

In 1986, Müller and Bednorz discovered a new class of superconducting materials, copper oxides, that superconducted at temperatures moderately higher than known before (30 K). Shortly thereafter, Chu showed that a small modification of the compounds discovered by Müller and Bednorz allowed superconductivity at temperatures substantially higher than the temperature where nitrogen becomes liquid, 77 K. This was the beginning of the high-temperature superconductivity 'revolution'. Today, compounds of that type superconduct till approximately 140 K (160 K under pressure). Ever since the cuprates were discovered, there was no doubt in anybody's mind that BCS theory cannot describe all superconductors: nobody doubts that cuprates are not described by BCS, but there is no agreement on what is the theory that describes them.

To me, the Bednorz, Müller, and Chu discovery was extremely important since it launched me in the research direction that I have been pursuing for the last 30 years, that this book is about. Before Bednorz and Müller, I was convinced, like all other physicists, that BCS was correct. Since 1988, never again.

Next, I would like to postulate as another hero of superconductivity somebody that nobody knows as related to superconductivity as of yet, but I hope some day will be recognized as such: Hannes Alfven. And not only because his name begins with *'H'*. Figure 2.5 shows his image, juxtaposed to that of another major hero of this story already mentioned, Werner Heisenberg, who introduced the concept of holes in solid state physics (in addition to inventing quantum mechanics). In contrast to Heisenberg, Hannes Alfven's work was exclusively in classical physics, no quantum physics was involved. To understand superconductivity, quantum mechanics at a macroscopic scale, we will need to fuse quantum and classical physics concepts.

Hannes Alfven was a theoretical plasma physicist. A plasma is a fluid of charged particles, in general positively and negatively charged particles in the same amount, that conducts electricity. There are also 'non-neutral plasmas', where the net charge is not zero. Alfven received the 1970 physics Nobel prize for his theoretical work on magnetohydrodynamics of plasmas. Coincidentally, Alfven was on the faculty at the same university I am at, UCSD, from 1967 until 1991 when he retired. But he was in the electrical engineering department, not in the physics department where I am at, and I never met him during the 9 years that we were colleagues. Furthermore, I didn't

Fig. 2.5 Left: Werner Heisenberg, one of the creators of quantum mechanics, introduced the concept of holes in solid state physics. Right: Hannes Alfven, 1970 physics Nobel laureate for his theoretical contributions to magnetohydrodynamics of plasmas. He formulated Alfven's theorem.

realize that Alfven's work had anything to do with superconductivity until several years after Alfven passed away in 1995.

Alfven never worked on superconductivity. But he is famous among other things for the so-called 'Alfven's theorem', formulated by him in 1942, which states:

> *In a perfectly conducting plasma, magnetic field lines behave as if they were moving with the plasma, i.e. as if they were frozen inside the plasma.*

In Chapter 10, I will explain why this concept is indispensable to understand the most fundamental property of superconductors that BCS should understand but doesn't: the Meissner effect.

Another reason why *superconductivity properly understood begins with H*: the theorem of *H*annes Alfven.

Finally, to conclude this chapter, I would like to return to its beginning, where we talked about Kammerlingh Onnes and helium liquefaction. I would like to mention here two discoveries by Onnes in liquid helium that are important in relation with superconductivity, although nobody knows it.

The first is something that Onnes discovered in 1911, coincidentally the same year that he discovered superconductivity. Onnes measured the density of liquid helium as a function of temperature, and discovered that on cooling helium its density increased to temperature 2.17 K, on further cooling its density decreased. Onnes found this behavior to be extremely interesting and peculiar, he returned to investigate it several years later, but could never explain it. Later in this book, I will explain something fundamental about the physics of superconductors that this discovery of Onnes illustrates, that BCS theory doesn't know.

The second discovery of Onnes in liquid helium I would like to mention is called the 'Onnes effect', discovered by him in 1922. It is the observation that liquid helium at low temperatures creeps up the walls of the container it is in, and flows out. This also illustrates physics of superconductivity, as I will explain later. This is also something that BCS theory doesn't know.

Superconductivity properly understood begins with Helium.

Figure 2.6 shows many of these landmarks and personalities in the development of superconductivity, including some not yet mentioned.

We have a long road ahead. Let's begin our journey.

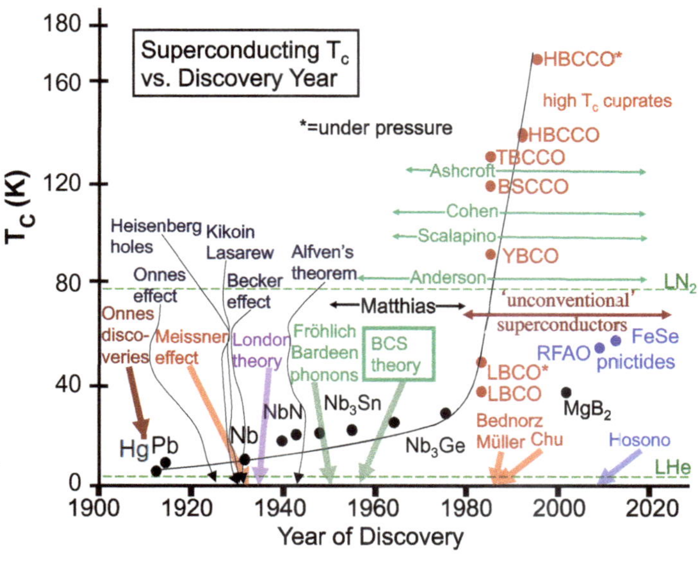

Fig. 2.6 Critical temperature of superconducting materials, effects, theories, and personalities in the history of superconductivity throughout the years. Some not yet mentioned in the text.

Chapter 3

Bird's eye view of superconducting materials

Superconducting materials come in three types: 'conventional', 'unconventional', and 'undetermined'. A compilation of superconducting materials divided into 32 different classes, 12 conventional, 11 unconventional, and 9 undetermined, can be found in [1]. Figure 3.1 shows some superconducting materials, their critical temperature, and the years of discovery.

'Conventional superconductors' are those materials about which there is general agreement that they are governed by the conventional theory of superconductivity, BCS. In particular, the electron–phonon interaction gives rise to pairing of electrons that in turn gives rise to superconductivity. They include essentially all superconducting materials known before 1980, as well as many others discovered after 1980. In particular, all superconducting elements (e.g. Hg, Pb, Nb) are considered conventional superconductors, as are binary compounds (e.g. Nb_3Ge) and alloys. It was believed that there could be no conventional superconductivity at temperatures above about 25 K. However, in 2001, the superconductor MgB_2, which is considered to be conventional and has critical temperature 39 K, was discovered.

'Unconventional superconductors' are those materials about which there is general consensus that they are NOT governed by conventional BCS theory. The most important ones are the class of copper oxide compounds (materials with CuO in Fig. 3.1, also known

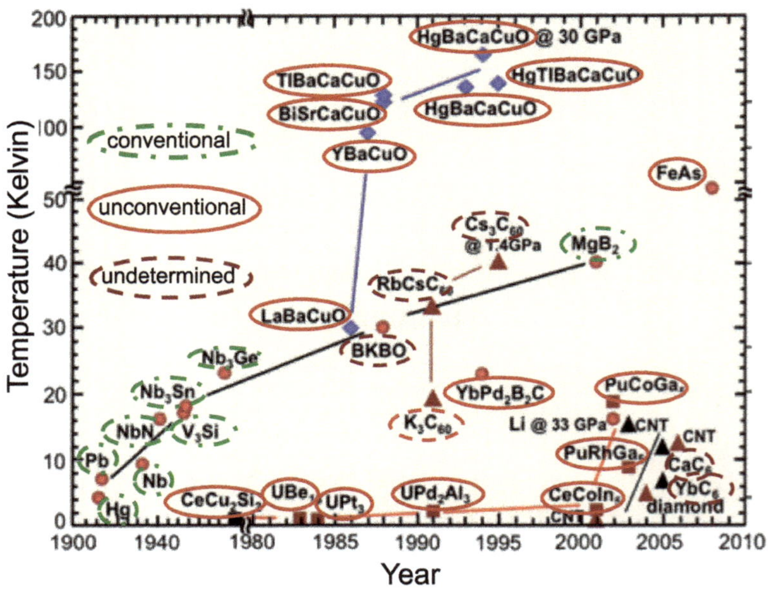

Fig. 3.1 Some superconducting materials, their critical temperature, and the years of discovery.

as cuprates) discovered in 1986, with critical temperatures that reach 160 K. In 2008, another important class of unconventional superconductors was discovered: arsenic iron compounds (FeAs) (also known as iron pnictides) with critical temperatures up to 60 K. Other examples of unconventional superconductors are 'heavy fermion materials' (UBe_{13}, UPt_3) and certain organic compounds.

'Undetermined superconductors' are those materials about which there is no general agreement on whether they are conventional or unconventional: some believe one, others the other. Examples are bismuthates (BKBO) and carbon compounds with the molecule C_{60} (Cs_3C_{60}).

The reader will note in Fig. 3.1 that for a material to be classified as 'unconventional', it is not necessary that it has a high critical temperature. Many superconductors considered to be unconventional have critical temperatures that are comparable or lower than that of many conventional ones, but have other characteristics that in appearance are not explained by BCS theory, for example, the temperature dependence of the specific heat. The general consensus is that conventional superconductors have a quantum wavefunction s, which means it is spatially isotropic, like a sphere, i.e. without angular dependence. Instead, unconventional superconductors have an anisotropic wavefunction, p or d, as Fig. 3.2 illustrates. It is believed that in unconventional superconductors the interaction responsible for pairing of electrons that leads to superconductivity originates in magnetic interactions between electrons and not in the electron–phonon interaction as BCS assumes.

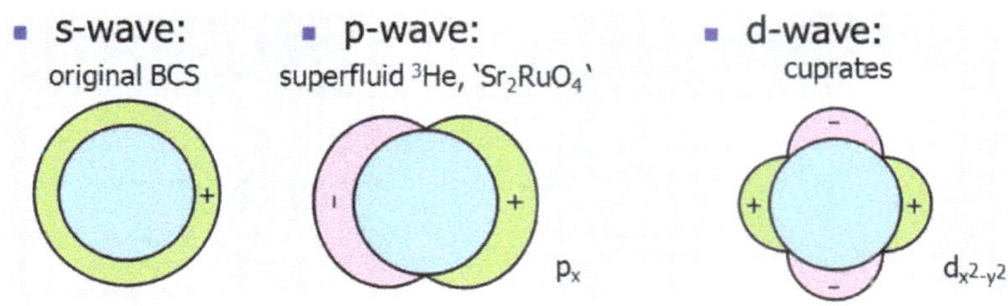

Fig. 3.2 Wavefunction of conventional superconductors (left) and of unconventional superconductors (center and right) according to the general consensus.

But this is theory. Couldn't reality be simpler than what Fig. 3.2 illustrates? Couldn't all superconducting materials be in essence the same? Couldn't the interaction that leads to superconductivity be always the same? After all, the most important characteristics of superconductors, that they conduct without electrical resistance and that they expel magnetic fields, are common to all. What if we assume superconductors are in essence all the same and that their differences in certain physical properties, including their T_c's, are due to details that are not essential to superconductivity? If that was the case, what Fig. 3.2 shows would be a figment of scientists' imagination, not reality.

If that was the case, we should identify some property of materials that is essential for them to be superconductors, which differentiates them from other materials that are not superconductors. All superconducting materials should have this property and non-superconducting materials should not have it.

In 1988, more precisely on October 28, 1988, at approximately 10 pm, I had an idea: perhaps what all superconducting materials have in common is that conduction is through holes rather than electrons. This occurred to me because I had just attended a scientific meeting on superconductivity in Japan, where it had been discussed that in the copper oxides, the high temperature superconductors discovered two years earlier, the charge carriers apparently were holes.

I will explain later what this means. For now, it is sufficient to know that the physical quantity one measures to determine whether conduction is through electrons or through holes is the Hall coefficient, R_H. When R_H is positive the charge carriers are holes, when R_H is negative they are electrons. Sometimes it is not unequivocal, there are materials with carriers of both types, electrons and holes, for which the Hall coefficient can be either negative or positive.

When this idea occurred to me, I was on an airplane flying at 10,000 m altitude over the Pacific returning from Japan to the US. I said to myself: (1) If this is so, it should be obvious in checking the Hall coefficient of many materials, that positive Hall coefficients predominate in superconductors and negative Hall coefficients in nonsuperconductors. I had never paid attention to this and had no idea what the answer would be. And (2): if this is so, for sure somebody must have realized this before. Unfortunately, there was no web connection in airplanes at that time (the internet didn't yet exist), so I had no way to check the literature. I waited anxiously for many hours unable to sleep until the plane landed.

Figure 3.3 shows the correlation between the sign of the Hall coefficient R_H and the existence or nonexistence of superconductivity in the elements. The correlation is extraordinary. The great majority of superconducting elements have positive R_H and the great majority of nonsuperconducting metallic elements have negative R_H. In 1997, I calculated that the probability that this would be due to chance, i.e. if it had nothing to do with the physics of superconductivity, as BCS theory says, is 0.001% (one in a hundred thousand), i.e. like buying a $10 lottery ticket and winning a million. Something similar occurs with compounds, i.e. materials with more than one element.

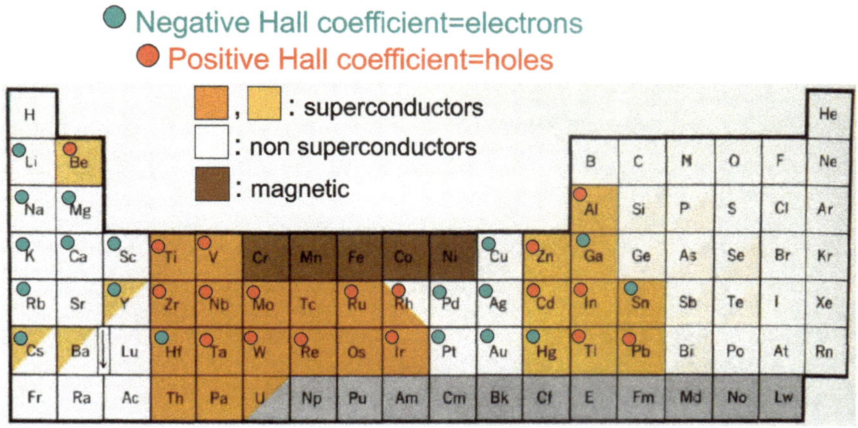

Fig. 3.3 Superconducting and nonsuperconducting elements and the sign of their Hall coefficient R_H.

I also verified immediately that my hypothesis (2) was correct. I found that a Russian physicist, I. Chapnik, had written papers in 1962, 1979, and 1983 pointing out this strange correlation between Hall coefficient and superconductivity [2] that did not have a theoretical explanation. In fact, he was not the first. In 1932, the Russian physicists Kikoin and Lasarew [3] pointed it out for the first time, and this was also discussed in papers by others, Papapetrou in 1934, Born and Cheng in 1948, and Feynman in 1956. Because BCS theory has nothing to say about this, it is not discussed in superconductivity textbooks and practically no contemporary physicist knows this. The Chapnik papers mentioned above have in total 48 citations, of which 28 are from papers I wrote and 7 are self-citations. The paper by Kikoin and Lasarew has 16 citations, 11 of which are from articles I wrote.

From that moment on in 1988 I have worked based on the conviction that *holes* are essential to superconductivity [4]. However, the fundamental reason why this is so I only understood on February 10, 2016. Since it took me almost 28 years to understand it, I ask the reader to be patient, I will explain it towards the end of this book. Like in detective novels, perhaps some readers will deduce it before reaching the end. I can anticipate that it is extremely simple.

Many parts in the development of this theory, particularly the first part, were done in collaboration with my colleague Frank Marsiglio, beginning when he was in San Diego as post-doctoral research associate in the period 1988–1991, and continuing until the present. His contribution to this work has been fundamental.

References

[1] Physica C Special Issue: Superconducting materials: Conventional, unconventional and undetermined, edited by J. E. Hirsch, M. B. Maple and F. Marsiglio, *Physica C* **514**, 1–444 (2015).
[2] I. M. Chapnik, *Sov. Phys. Dokl.* **6**, 988 (1962); On the empirical correlation between the superconducting T_c and the Hall coefficient, *Phys. Lett. A* **72**, 255 (1979); *J. Phys. F* **13**, 975 (1983).
[3] K. Kikoin and B. Lasarew, Hall effect and superconductivity, *Nature* **129**, 57–58 (1932).
[4] J. E. Hirsch, Hole superconductivity, *Phys. Lett. A* **134**, 451 (1989).

Chapter 4

BCS and Hubbard: Theories of superconductivity that don't explain it

In the mind of physicists working on superconductivity, it is firmly established that BCS theory with the electron–phonon interaction (Fig. 4.1 shows it schematically) describes conventional superconductors. It is not something that can be examined or questioned. Nor it is something that can be experimentally disproved. If a material does not adjust in some aspect to what is expected according to BCS theory, first it is tried to make the calculation more complicated, e.g., extending the model so it has more adjustable parameters, in order to fit the experiments. If it is impossible, the material is simply declared to be 'unconventional', i.e. not described by BCS.

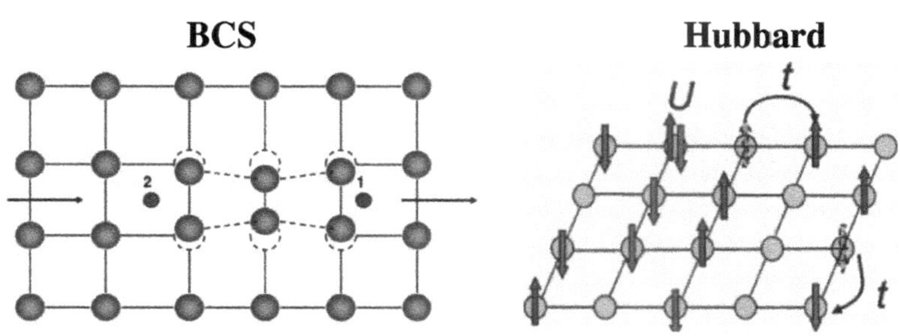

Fig. 4.1 Microscopic physics of BCS theory (left) and Hubbard model (right). In BCS, electrons pair due to an effective attraction that results from the ionic displacement. In Hubbard, electrons repel strongly with interaction energy U when they are on the same site.

How can science advance with this attitude? It is a religious attitude, no data from reality can affect a theological belief.

Note that in the early days of superconductivity, before BCS, scientists were searching for *the* explanation of superconductivity. That is, a *single* explanation that would apply to all superconducting materials. It is a logical stance. Superconductivity is such a unique phenomenon, so different from the typical behavior of materials at the macroscopic scale, that it is illogical to think there may be three or five or ten different explanations for it, with different symmetries as we saw in Fig. 3.2. It is more reasonable to expect that there is a unique mechanism and a unique symmetry, and if there are features that appear distinct in different materials, they are accessory details that are irrelevant to the essence of what determines that there is superconductivity in the material.

Immediately after BCS, 1957, this continued to be the attitude of the scientific community, and when a material showed a discrepancy with the predictions of BCS theory, some other physics was invoked to explain the discrepancy. For example, if T_c was found to be much lower than expected within BCS, or even zero, it was argued that the material had 'spin fluctuations' that suppressed BCS superconductivity. Or, that electrons had a larger Coulomb repulsion in those materials. But eventually that attitude became unsustainable, particularly after the discovery of the cuprates in 1986, and it was necessary to accept that BCS with the electron–phonon interaction cannot explain the superconductivity of all superconducting materials. Nobody doubts that today.

But then, when a new material is discovered that doesn't conform to BCS, rather than asking 'could it be that BCS is not right?', scientists say 'this material is *unconventional*, BCS doesn't apply to it'. Excellent excuse. In other words, BCS theory is not falsifiable by finding materials that do not conform to it: the theory is not wrong, the material is wrong.

Typical (verbatim) comments that I get from referees when I write articles that develop physics of superconductivity that is not part of BCS, hence question BCS:

- BCS theory is one of the most successful microscopic theories of condensed matter physics.
- BCS theory is universally accepted as one of the greatest, if not the greatest achievement in modern condensed matter physics.
- The BCS–Eliashberg theory continues to be universally accepted as the crowning achievement of condensed matter physics and he is alone in opposition.
- The idea that all superconductors share the same mechanism is laughable on the face of it, since we find both triplet and singlet cases and several order parameter symmetries, as well as real and complex gaps, and each case needs a new motivation.
- There is no need for a new explanation of Meissner effect that is based upon the 19th century physics.

Etcetera. Based on this, the referees don't consider my specific arguments when I propose different physics, they don't contradict my arguments with physical arguments, they simply ignore them.

In the year 2007, a conference was held at the University of Illinois to commemorate the 50th anniversary of BCS theory. Participants included Cooper and Schrieffer of BCS (Bardeen passed away in 1991), several of the most prominent condensed matter theorists and experimentalists working in the area of superconductivity, both 'conventional' and 'unconventional', and even very prominent high energy physicists. The proceedings of the conference were published by World Scientific [1]. In the midst of the 'unconventional superconductivity' era, there was not a single suggestion in the entire conference of the possibility that BCS might not be completely correct in its description of 'conventional' superconductivity.

After the discovery of the cuprates in 1986, once it was accepted by the scientific community that the BCS religion cannot apply to all superconducting materials, the 'strongly correlated electrons'-Hubbard model cult was born.

The Hubbard model was proposed by British physicist John Hubbard in 1963 to try to explain the phenomenon of ferromagnetism in solids. It is a very simplified model of reality. It is assumed that there is a single atomic orbital per atom for conduction electrons to reside in, and that two electrons interact repulsively with each other only when they are on the same atom and hence in the same orbital. These are enormous simplifications, since a real atom has an infinite number of atomic orbitals, not one, and in addition the Coulomb interaction between electrons is not local, on the contrary, it decays slowly with distance r as $1/r$. Even so, the great majority of theoretical physicists that want to understand 'correlated electrons' and the physics of high temperature superconductivity concentrate on studying the Hubbard model. How do they justify this? They reason (correctly) that if the BCS electron–phonon interaction cannot explain high temperature superconductivity, it must result from the electron–electron interaction. Then, they say that the Hubbard model is 'the simplest model that describes interacting electrons', so one has to understand it first, once we understand it we can try to understand more realistic models with more extended interactions and more than one orbital per atom.

But this argument is false. The Hubbard model, even though it looks simple, is not exactly solvable and it is very difficult to establish its behavior through approximate analytic or numerical methods. Despite the enormous efforts devoted to it over the past 30 years, there is no agreement whether the Hubbard model exhibits superconductivity at any temperature or not. In my opinion, and based on calculations that I performed more than 30 years ago, no. But more importantly, in my opinion that is completely irrelevant. There is no reason to assume that the Hubbard model has anything to do with the physics of the cuprates, in particular, with the physics of cuprates that is responsible for their superconductivity at high temperatures.

The scientific community has a different opinion. The prestigious journal *Nature* published an article in 2013 [2], 'The Hubbard model at half a century', describing the enormous popularity of the model in solid state physics over the years up to today.

What is the explanation for this Hubbard model 'cult'? It's not completely clear to me. I think it is part of a general preference of solid state theorists to study 'models' instead of studying physical systems. Somebody famous proposes a model (for example,

Phil Anderson has proposed several, or John Hubbard, or Jun Kondo, or Alexei Kitaev) and a multitude of young physicists with higher analytic and/or numerical skills than the more mature physicists launch efforts to study the 'Hubbard model', or the 'Anderson impurity model', or the 'extended Anderson model', or the 'Kondo model', or the 'Kitaev model'. They publish article after article about the properties of these models, totally forgetting to ask whether the models have anything to do with the real physics of real materials. It is even regarded as somewhat demeaning to worry about details of real materials as opposed to concentrating on the 'pure' physics of these idealized models.

The result? Progress over the last 30 years in understanding the origin of superconductivity in the cuprates has been nil. The same with other 'unconventional' superconductors such as the iron–nitrogen compounds (iron pnictides) discovered in 2008. My explanation is the following: all physicists trying to understand 'unconventional' superconductivity using the Hubbard model, or other models, start from the assumption that conventional superconductors are correctly and completely explained by BCS theory. Explicitly or implicitly, this entirely conditions their reasoning. Because of this, it is impossible that they make progress. The first step to make progress in understanding the so-called 'unconventional superconductivity' is, in my opinion, to accept the possibility that so-called 'conventional superconductivity' is not well understood by the conventional theory, BCS.

Figure 4.1 shows schematically the microscopic physics involved in BCS theory and Hubbard models. In BCS, the displacement of the ions is the essential element that leads to pairing of electrons, the Coulomb interaction between electrons plays no role. In Hubbard, the displacement of ions plays no role, the repulsive interaction U between electrons is the essential element, despite being repulsive electrons magically pair. These are completely different microscopic physics, in some sense diametrically opposed, which according to the general consensus give rise to the same macroscopic phenomenon, superconductivity. One 'conventional', the other 'unconventional'.

A famous phrase of Barack Obama, recent US President:

There is not a liberal America and a conservative America — there is the United States of America. There is not a black America and a white America and Latino America and Asian America — there's the United States of America.

I would like to paraphrase it as follows:

There is no conventional and unconventional superconductivity — there is no singlet and triplet superconductivity — there is no s, p and d superconductivity — there is no superconductivity with real and complex order parameters — there is no topological superconductivity and exotic superconductivity — there's superconductivity.

What none of these physicists working on these diverse, imaginative, and fanciful attempts to understand superconductivity take into account is what is called 'Ockham's razor': the philosophical principle that states that simple and general explanations should be preferred over complicated and particular explanations.

They also don't take into account what I consider to be the most fundamental element of *all* superconductors, indispensable to understand superconductivity: *charge asymmetry.* We start with that subject in Chapter 5.

References

[1] *BCS: 50 Years*, ed. by L. N. Cooper and D. Feldman, World Scientific, Singapore (2010).

[2] The Hubbard model at half a century, *Nature Physics* **9**, 523 (2013).

Chapter 5

Electrons and holes in solids: The key to superconductivity

The concept of holes is simple. Consider a box that can fit exactly N objects. If the box has n objects, with n smaller than N, we can say that the box has n objects or that it has N–n holes. Suppose the objects are electrons, that have negative charge, and that there is a background of positive charge that neutralizes it, for example at the box's walls. We can then say that the absence of an electron, the hole, has positive charge of the same magnitude.

Werner Heisenberg introduced the concept of holes in 1931 [1], first to explain optical spectra of atoms and molecules. In atoms there are shells that can hold eight electrons. For example, the argon atom (Ar) has eight electrons in its outermost shell, it is full. The chlorine atom (Cl), immediately to the left in the periodic table (Fig. 3.3) has seven electrons, or equivalently one hole.

Similarly as the atoms, solids also have 'shells', that are denoted by 'bands'. When a band is full of electrons, and there are no other bands close in energy, electrons have nowhere to move and the solid is an insulator, it cannot conduct electricity. When the band is partially full, there is room for electrons to move to states of similar energy with finite momentum, and the solid conducts electricity, it is a metal. If a band can fit N electrons and has n electrons, we can say that electric current is carried by n electrons of negative charge or by (N–n) holes of positive charge. For a given direction of the current, electrons move in opposite direction to holes. Holes move in the

Fig. 5.1 The hole is denoted by the dotted circle. It moves from the second to the third place in the row, i.e. to the right. Equivalently, an electron moved to the left, from third to second place.

direction of the current, electrons move in direction opposite to the current. In Fig. 5.1, we can say that an electron moved from right to left, or equivalently that a hole moved from left to right.

It is important to clarify that we shouldn't speak of electrons and holes at the same time, we need to decide. Either we talk about electrons or we talk about holes. How do we decide? Very simple: if the band is less than half full, there is fewer electrons than holes, then we talk about electrons. If it is more than half full there are fewer holes, we talk about holes and we forget about electrons.

Then is it a question of language that we can decide by this convention or another one? Not really. There is a physical quantity that we can measure in the laboratory that tells us whether the band is filled more or less than half, and consequently whether the charge carriers are holes or electrons. The Hall coefficient previously mentioned, denoted by R_H.

To understand the Hall coefficient, we need to remember the formula for the force that electric and magnetic fields exert on charges, the so-called Lorentz force. The formula is

$$\vec{F} = q\vec{E} + q\frac{\vec{v}}{c} \times \vec{B} \equiv \vec{F}_E + \vec{F}_B \tag{5.1}$$

In this formula, q is the electric charge of a particle, $\vec{E}$ and $\vec{B}$ are electric and magnetic fields, $\vec{v}$ is the velocity of the particle, and c is the speed of light, 300,000 km/s.

Figure 5.2 shows what happens in a 'Hall bar' used by Edwin Hall, a doctoral student at John Hopkins university in 1879 when he discovered the effect that bears his name, the Hall effect. J_x denotes the density of electric current that flows in the x direction. The charge carriers move in direction x if they are positive holes (b), or in direction -x if they are negative electrons (a). B denotes the magnetic field in the vertical direction (z). The magnetic Lorentz force $\vec{F}_B$

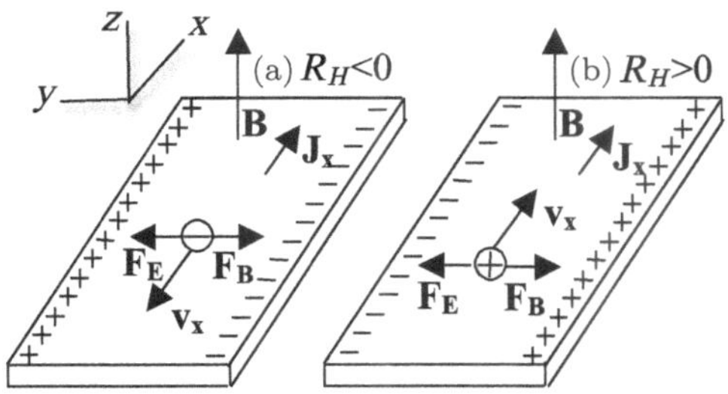

Fig. 5.2 Electrical conduction in materials where the charge carriers are electrons (a) and holes (b). The transverse voltage that gets generated (in the y direction) has opposite sign for both cases. Measuring this voltage we determine whether the charge carriers are electrons or holes and their density.

(Eq. (5.1)) is to the right both for electrons and for holes, as the reader can verify using the right-hand rule for the vector product in formula (1), taking into account that q is negative for electrons and positive for holes. This gives rise to an accumulation of charge with the signs shown in the figure, that is detected experimentally by connecting a voltmeter to both sides of the bar. The Hall coefficient thus measured, denoted by R_H, is proportional to that voltage, and its sign is negative when the charge carriers are electrons (a) and positive when they are holes (b).

The amount of charge accumulated on the lateral walls is such that the electric field generated by this charge exerts a force F_E on the charge carriers in the y direction that cancels the magnetic force F_B, so that both holes and electrons move in a straight line, in direction x or $-x$.

Performing this experiment, Hall found that, for example, for the metals lead (Pb), niobium (Nb), and tin (Sn) R_H is positive, and for copper (Cu), silver (Ag), and gold (Au) it is negative. He didn't know at the time, since superconductivity had not yet been discovered, that Pb, Nb, and Sn are superconductors at low temperatures and Cu, Ag, and Au are not.

If the reader is satisfied with the explanation I just gave for the sign of R_H, then either he/she knows a lot of physics and can plug the 'holes' in my explanation, or is not paying much attention. Because the explanation I gave doesn't make much sense. I said we can think of the motion of charge equivalently as negative charge (electrons) moving in one direction or positive charge (holes) moving in the opposite direction. If we think about it always with electrons, we would obtain always what Fig. 5.2(a) shows, corresponding to negative R_H. How come sometimes R_H is positive?

When Edwin Hall measured this effect in 1879, he expected to find only negative values of R_H, when he found positive values in some metals like Pb he couldn't explain this, and this was known for many years as the 'anomalous Hall effect'. It was only explained in 1929 by a young German physicist, Rudolph Peierls, Ph.D. student of Heisenberg, using the concept of holes suggested to him by Heisenberg [2].

The explanation that Peierls gave is based on the concepts introduced by Felix Bloch (another German physicist) in 1928 to describe electrical conduction in solids. We will not go into details here. The essence is that one has to take into account the forces exerted between the electrons and the ions in the solid when there is electrical conduction, in addition to the forces exerted by the external electric and magnetic fields given by Eq. (5.1). When all the forces are taken into account, Peierls showed that for cases where the band is less than half full the behavior is as if the carriers were electrons, that is Fig. 5.2(a), and when the band is more than half full the behavior is as if the carriers were holes, that is Fig. 5.2(b). In other words, the result is intuitive and looks simple, but the proof is not.

Generally in physics books, the explanation is given showing Fig. 5.2, without entering into detail. For that reason, many physicists are not familiar with the correct explanation, they think about 'holes' as if they were real particles, as electrons are, when in reality they are not. They believe there is an 'electron–hole symmetry' in condensed matter. I myself was confused about this point, and it took me 28 years to realize that understanding correctly Fig. 5.2 we understand an absolutely essential point about superconductivity. We will return to this later in this book.

The reality is, there exists a fundamental *A*symmetry between electrons and holes. The origin of this asymmetry is that in nature negative charges (electrons) and positive charges (protons) do NOT have the same mass, but rather have masses that differ by a factor 2000. This fundamental charge asymmetry is what makes superconductivity possible, as we will see in this book, contrary to what the generally accepted theory says.

References

[1] W. Heisenberg, Zum Paulischen Ausschliessungsprinzip, *Annalen der Physik* **402**, 888 (1931).
[2] R. Peierls, Zur Theorie des Hall-Effekts, *Physikalische Zeitschrift* **30**, 273 (1929); Zur Theorie der galvanomagnetischen Effekte, *Z. für Physik* **53**, 255 (1929).

Part II

THE KEY QUESTIONS THAT NEED TO BE ANSWERED

Chapter 6

The simplest question in superconductivity, that BCS doesn't answer

Before beginning the discussion of superconductivity theories, I would like to pose an apparently trivial question about superconductivity: how does the current in a superconductor stop?

Consider Fig. 6.1. On the left panel, there is a superconducting cylinder to which a magnetic field B will be applied. When B is applied, an electric current I in clockwise direction starts flowing near the surface of the cylinder, corresponding to electrons moving counterclockwise. This current I generates a magnetic field in the interior of the cylinder that points in direction opposite to the applied field B, i.e. down. Added to the applied field B, it yields zero for the total field in the interior of the cylinder, as shown in the center panel of Fig. 6.1. As explained earlier, superconductors do not tolerate magnetic fields in their interior.

To understand this process, that is from the left to the center panel of Fig. 6.1, it is sufficient to know Faraday's law, formulated mathematically by the British physicist James Clerk Maxwell in 1861. It states that when a magnetic field B changes in time an electric field E_F, Faraday's field, is induced in direction such that it induces an electric current that itself generates a magnetic field opposite to the change in the external magnetic field. In other words, on the left panel there is no magnetic field in the interior of the superconductor. If we try to introduce it, Faraday generates the electric field E_F (center panel) that propels the current I, that generates a

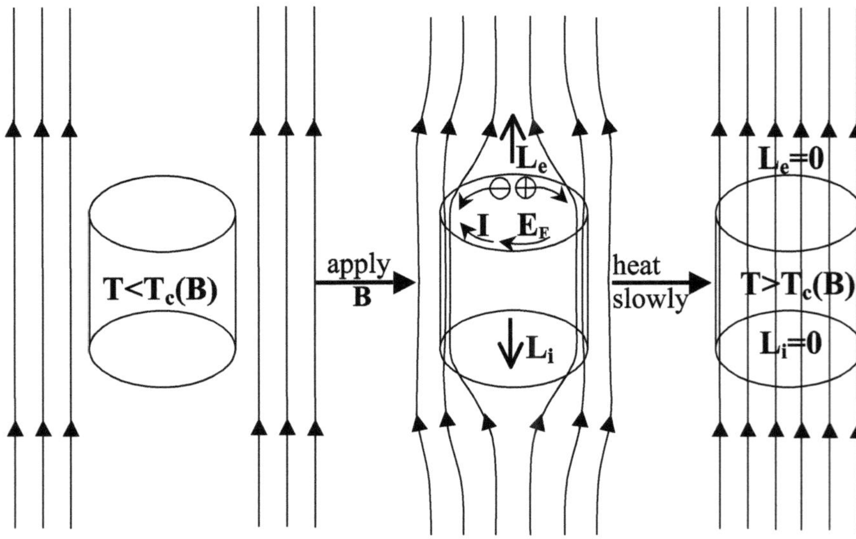

Fig. 6.1 When a magnetic field is applied to a superconductor (left panel), a surface electric current I is generated that prevents the field from entering the superconductor. If we then raise the temperature, the superconductor becomes normal, the current stops, and the magnetic field enters the body (right panel). How this occurs is a mystery.

field $-B$, keeping the interior of the material without a net magnetic field, as the central panel in Fig. 6.1 shows. It is as if a barrier exists on the surface of the cylinder that prevents magnetic field lines from penetrating. If the material was not a superconductor, the electric current induced by the Faraday field would decay with time and eventually disappear, then the field B would penetrate the cylinder.

In the superconductor, the electric current persists while the material is superconducting, and the field B does not penetrate. What happens if we then heat the superconductor to a temperature above its critical temperature, and it becomes a normal metal? The current will stop flowing in the normal metal since there is no battery to keep it going. The magnetic field penetrates the cylinder, as shown in the right panel of Fig. 6.1.

Question: How does the current stop in the process starting from the central panel of Fig. 6.1 and ending in the right panel?

The intuitive answer is: When the superconductor becomes normal, the resistivity becomes nonzero. The current stops by dissipating its energy as heat. Microscopically, the charge carriers collide with the ions in the solid, they lose their kinetic energy and their mechanical momentum, and they stop.

The problem is, that answer is wrong. That was what was believed to be true until the year 1933, when the Meissner effect was discovered.

The Meissner effect is the inverse of the process between the center panel and the right panel in Fig. 6.1. If we start with the normal metal in a magnetic field (right panel) and lower the temperature, the system evolves to the central panel. A current I is spontaneously generated that generates a magnetic field opposed to the applied field B, and the result is the central panel: the magnetic field was expelled from the interior of the superconductor.

We will discuss the Meissner effect in Chapter 7. Here I only want to point out: because the transition between the center panel and the right panel in Fig. 6.1 can happen in both directions, i.e. from left to right or from right to left, it is what is called a *reversible process.* Thermodynamics teaches us that in a reversible process energy cannot be dissipated. So, in going from the center to the right panel, how does the current stop without dissipating energy? It is a basic question that, incredibly, was never asked by theoretical physicists working on superconductivity, no matter how high their H-index. And it is a question that I say the accepted theory of superconductivity, BCS, cannot answer.

There was however an experimental physicist that did ask himself this question. Willem Hendrik Keesom, a disciple of Kammerlingh Onnes, made very important experimental contributions to low temperature physics, liquid helium, and superconductivity. In the years 1934–1938, he performed with his collaborators a series of measurements [1] trying to detect whether heat was dissipated in an irreversible way in the transition superconductor-to-metal shown in Fig. 6.1. He showed experimentally that certainly not more than 1% of the kinetic energy of the supercurrent was dissipated, which strongly suggests that none is dissipated. From this, he deduced that *'it is essential that the persistent currents have been annihilated before the material gets resistance, so that no Joule-heat is developed'*. But he did not propose an explanation for how this happens, and the physicists that followed him did not pay any attention to the important question that Keesom had formulated and investigated but not answered.

What is relatively simple is to explain mathematically how the current in the central panel of Fig. 6.1 is generated starting from the left panel. Faraday's law expressed mathematically in differential form is

$$\vec{\nabla} \times \vec{E} = -\frac{1}{c}\frac{\partial \vec{B}}{\partial t} \tag{6.1a}$$

and in integral form

$$\oint \vec{E} \cdot d\vec{l} = -\frac{1}{c}\frac{\partial}{\partial t}\int \vec{B} \cdot d\vec{S} \equiv -\frac{1}{c}\frac{\partial}{\partial t}\phi \tag{6.1b}$$

where c is the speed of light and ϕ is the magnetic flux across a surface S. The line integral on the left side of (6.1b) is along a contour that borders S. The electric field E in this equation is Faraday's field E_F. On the other hand, we also will need Ampere's law, that states

$$\vec{\nabla} \times \vec{B} = \frac{4\pi}{c}\vec{J} \tag{6.2}$$

where $\vec{J}$ is the electric current density, given by

$$\vec{J} = n_s e \vec{v}_s \tag{6.3}$$

where n_s is the density of superconducting electrons (number of electrons per unit volume) and $\vec{v}_s$ is the velocity of the supercurrent.

The equation that describes the current generation on the central panel of Fig. 6.1 is

$$m_e \frac{\partial \vec{v}_s}{\partial t} = e\vec{E} \tag{6.4}$$

m_e is the electron mass and $\vec{E}$ is the electric field. Equation (6.4) is Newton's law, with the electric Lorentz force given in Eq. (5.1).

Taking the curl on both sides of Eq. (6.4) and using Eq. (6.1a)

$$\frac{\partial}{\partial t}\vec{\nabla} \times \vec{v}_s = -\frac{e}{m_e c}\frac{\partial \vec{B}}{\partial t} \tag{6.5}$$

and integrating Eq. (6.5) in time,

$$\vec{\nabla} \times (\vec{v}_s(\vec{r}, t) - \vec{v}_s(\vec{r}, 0)) = -\frac{e}{m_e c}(\vec{B}(\vec{r}, t) - \vec{B}(\vec{r}, 0)). \tag{6.6}$$

At time $t = 0$, both the velocity and B are zero at any point $\vec{r}$ of the cylinder, so that

$$\vec{\nabla} \times (\vec{v}_s(\vec{r}, t)) = -\frac{e}{m_e c}\vec{B}(\vec{r}, t). \tag{6.7}$$

From now on, we will no longer write the variable $\vec{r}$, we leave it implicit. Using Eq. (6.3), we obtain from Eq. (6.7)

$$\vec{\nabla} \times \vec{J} = -\frac{n_s e^2}{m_e c}\vec{B}. \tag{6.8}$$

If we now apply the curl operator to Eq. (6.2), use Eq. (6.8) for the right-hand side and use the vector relation

$$\vec{\nabla} \times \vec{\nabla} \times \vec{B} = -\nabla^2 \vec{B} \tag{6.9}$$

we obtain

$$\nabla^2 \vec{B} = \frac{1}{\lambda_L^2}\vec{B} \tag{6.10a}$$

with

$$\frac{1}{\lambda_L^2} = \frac{4\pi n_s e^2}{m_e c^2} \tag{6.10b}$$

and similarly obtain

$$\nabla^2 \vec{J} = \frac{1}{\lambda_L^2}\vec{J}. \tag{6.11}$$

These equations indicate that both the field B and the current J decay exponentially to zero over a distance λ_L from the surface of the superconductor, since they admit a solution

$$B(r) = Be^{(r-R)/\lambda_L} \tag{6.12a}$$

$$J(r) = Je^{(r-R)/\lambda_L} \tag{6.12b}$$

where R is the cylinder radius and r is the cylindrical radial coordinate. λ_L is called the London penetration depth, and is defined by Eq. (6.10b). It is of order 400 Å or somewhat larger. Remember that $1\,\text{Å} = 10^{-10}\,\text{m} = 0.1\,\text{nm}$.

The speed of the carriers of the supercurrent, v_s, can be obtained from Eq. (6.7). It is

$$v_s = -\frac{e\lambda_L}{m_e c}B. \tag{6.13}$$

For a typical magnetic field $B = 300\,\text{G}$ (G = Gauss), and $\lambda_L = 400\,\text{Å}$, the speed of the supercurrent carriers is

$$v_s = 21,075\,\text{cm/s} \tag{6.14}$$

much larger than the speed of charge carriers in a normal metal, which is of order mm/s. These carriers moving at high speed have to stop when the system becomes normal in the right panel, through an unknown process that does not dissipate energy. The kinetic energy of the current needs to be stored somewhere, to be available if we want to do the process in the opposite direction, in the Meissner effect that we will analyze in Chapter 7. The kinetic momentum of the charge carriers also has to be 'stored' somewhere, due to the principle of momentum conservation.

BCS theory does explain where the kinetic energy of the charge carriers is 'stored' when the superconductor becomes normal and the supercurrent stops: it is used to pay the difference in energies between the normal and the superconducting states, the normal state has higher energy. But BCS cannot explain where and how the momentum of the charge carriers is stored when the supercurrent stops.

Reference

[1] W. H. Keesom and J. A. Kok, "Measurements of the latent heat of thallium connected with the transition, in a constant external magnetic field, from the supraconductive to the non-supraconductive state", *Physica* **1**, 503 (1934); "Further calorimetric experiments on thallium", *Physica* **1**, 595 (1934); W.H. Keesom and P.H. van Laer, "Measurements of the latent heat of tin while passing from the superconductive to the non-superconductive state at constant

temperature", *Physica* **4**, 487 (1937); "Measurements of the atomic heats of tin in the superconductive and in the non-superconductive state", *Physica* **5**, 193 (1938); P.H. van Laer and W.H. Keesom, "On the reversibility of the transition process between the superconductive and the normal state", *Physica* **5**, 993 (1938).

Chapter 7

Meissner effect and London theory

To make progress in the understanding of superconductivity, BCS theory, and the alternative hole theory, we need to understand in some detail the Meissner effect and the phenomenological explanation proposed by the London brothers.

As we discussed in Chapter 6, Faraday's law says that a change in the magnetic flux induces an electric field (called "Faraday's field") that generates an electric current that in turn generates a magnetic field that *opposes* the change in magnetic flux.

We need to add to this: the current generated by the Faraday field depends on the conductivity of the material. If the material is an insulator (or if there is no material, rather empty space), the current is zero, the Faraday field exists but does nothing. If the material is a perfect conductor, the current is maximum (it is the case discussed in Chapter 6). The magnetic field generated by the current is zero in the first case (obviously), in the second case it is such that it completely cancels the initial change in magnetic flux, hence the magnetic flux doesn't change. In other words, the magnetic flux in a perfect conductor is 'frozen', whatever it is it cannot change (in Chapter 6, it was zero in the left panel of Fig. 6.1 and because of Faraday's law it stayed zero in the central panel). This is related to Alfven's theorem that we mentioned earlier and to which we will return later in the book.

This tells us that if a metal with a magnetic field inside becomes a perfect conductor when the temperature is lowered, the magnetic field should stay inside, frozen. Superconductors do exactly the opposite!

For a normal metal, i.e. intermediate between an insulator and a perfect conductor, the Faraday field gives rise to a current and resulting magnetic field that reduces the initial change in magnetic flux but doesn't cancel it completely. With time this current decays due to resistance, and the magnetic field in the interior can change.

In summary, Faraday's law doesn't like *changes* in magnetic field, it opposes them any way it can. It is equivalent in politics to an ultra-conservative person. Let's consider an example.

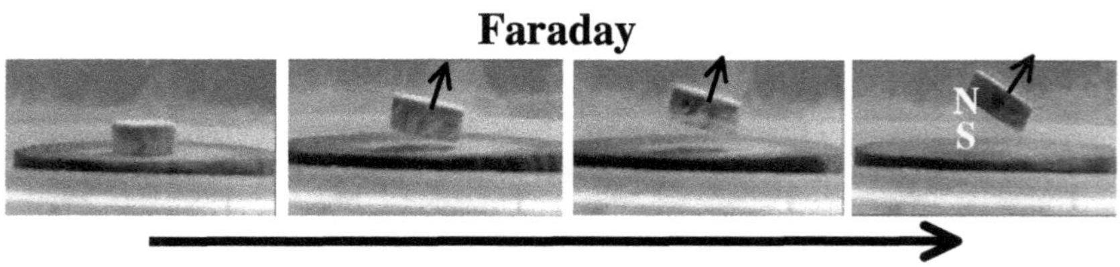

Fig. 7.1 Illustration of Faraday's law. The horizontal arrow indicates the direction of the process. The vertical arrows indicate an external force that we apply to lift a magnet that rests initially on top of a normal metal. The motion of the magnet converts the metal into a magnet that attracts the other magnet, making it harder for us to lift it. In other words, the metal resists the magnet being lifted.

In Fig. 7.1 left panel, a magnet rests on a metal disk. Moving to the right, we apply a force to the magnet, indicated by the vertical arrow, to lift it. We can imagine the vertical arrow is our hand pulling the magnet up. As a consequence, the magnet moves up. At the same time, an electric current appears in the metal, induced by Faraday, that converts the metal into another magnet of opposite polarity to the magnet above it. Hence, there is a south (S) magnetic pole on top of the metal that attracts the north (N) magnetic pole of the magnet above it, and makes it harder for us to lift the magnet. The force we need to apply to lift the magnet needs to overcome not only gravity but also the force of attraction between the magnet and the metal produced by Faraday's law. If the metal was a perfect conductor, this force of attraction would be infinite and we would not be able to lift the magnet, or, we would lift the metal together with the magnet, it would be as if the metal was 'glued' to the magnet.

The process also works in reverse. If we approach a magnet to a metal, the metal becomes a magnet with polarity such that it repels the magnet that is approaching, making the process more difficult. Faraday's law opposes whatever we want to do, no matter in which direction.

In contrast, let's see what happens in the Meissner effect [1].

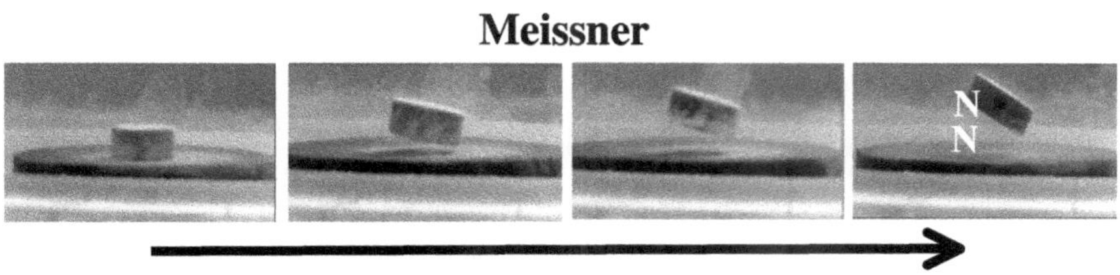

Fig. 7.2 Illustration of the Meissner effect. The horizontal arrow indicates the direction of the process. Here there is no external force applied, rather the metal is cooled and becomes superconducting. In the process, the metal becomes a magnet that *repels* the magnet on top of it and pushes it up. These are real images of a movie taken immediately after adding liquid nitrogen to the container shown in the figure.

In the Meissner effect, we have initially the same situation as before (Fig. 7.2 left panel). Now we don't apply an external force, what we do is we cool the metal to the point where it becomes a superconductor. What is observed, as the figure shows, is that the magnet moves up by itself, without any applied external force! In order for this to happen, the metal in the process of becoming superconducting has to itself become a magnet, but of opposite polarity from what we had in Fig. 7.1. Here, the current in the superconductor circulates in opposite direction to that in Fig. 7.1, generating a north pole on the top of the superconductor that repels the north pole of the magnet above it, pushing it up.

On the other hand, we can ask: what happens if we do the opposite process, as discussed earlier, that is we approach a magnet to a superconductor? The answer is, the same as a normal metal or a perfect conductor would do: the superconductor becomes a magnet of the same polarity as the one we are approaching so that it repels it, following Faraday's law.

So in one case the superconductor follows Faraday's law, in the other case, the superconductor does exactly the opposite of what Faraday's law dictates. How is that possible?

That is the mystery of the Meissner effect.

Going back to Fig. 7.2, it is important to remember that Faraday's law always acts. In the process where the magnet is lifted, Faraday's law creates an electric field in direction such that it wants to create an electric current that converts the superconductor into a magnet of polarity shown in Fig. 7.1, so that it attracts the magnet. But the superconductor does exactly the opposite. That is, it generates a current in opposite direction to what Faraday's law is telling it to do.

Let us now consider the same physics with diagrams that illustrate more clearly what's going on.

As Fig. 7.3 shows, the Meissner effect is the inverse of the process from the center to the right panel in Fig. 6.1. In Fig. 7.3, the cylinder

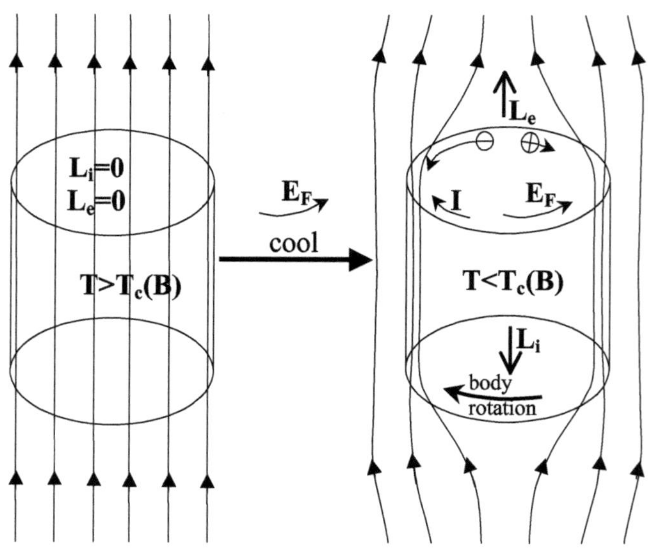

Fig. 7.3 Meissner effect: a normal metal in the presence of a magnetic field is cooled and becomes superconducting. In this process, an electric current I is spontaneously generated that nullifies the applied magnetic field in the interior of the material. Note that the Faraday field E_F that is generated as the magnetic field changes is in the direction of *stopping* the current I.

is initially in the normal state, with magnetic field in the interior pointing up. When it is cooled, a clockwise current I is generated that generates a magnetic field pointing down, and the magnetic field is expelled. In other words, the generated magnetic field added to the applied magnetic field add up to zero in the interior of the cylinder. This is essentially the same as we saw in Fig. 7.2. The cylinder here is the metal disk in Fig. 7.2, the magnet in Fig. 7.2 produces the magnetic field lines that we see in Fig. 7.3. When the magnet moves up in Fig. 7.2, the magnetic field lines are expelled as Fig. 7.3 shows.

This process is completely opposite to what we would expect according to the laws of classical physics, for the following reasons:

(1) Law of inertia: it says that a particle will remain in a state of rest if no force acts on it. In the left panel of Fig. 7.3, there is no current, hence electrons are (on average) at rest. When the system is cooled, electrons near the surface spontaneously start to move in counterclockwise direction, to expel the magnetic field. They don't do it if there is no applied magnetic field. *What is the force that propels the electrons to move?* The magnetic Lorentz force, Eq. (5.1), acts perpendicular to the magnetic field and to the velocity, and is proportional to the velocity. Here there is no initial velocity, hence the Lorentz force doesn't act. There is also no change in magnetic field produced externally, so there is no Faraday field initially. Why do electrons start moving?
(2) As the current I starts to flow, a magnetic field pointing downwards starts growing, opposing this a Faraday electric field in counterclockwise direction is generated that pushes electrons in clockwise direction, that is it opposes their motion. In other words, in generating the Meissner current, the electrons not only defy inertia, but more than that, they have to overcome the Faraday force that opposes their motion. The same that we said happened in Fig. 7.2.
(3) The law of momentum conservation says that the cylinder has to start rotating in direction opposite to the motion of the electrons, so total momentum is conserved. The Faraday field exerts a counterclockwise force on the positive ions, but the cylinder

has to rotate clockwise to compensate the momentum of the electrons. What is the force that makes the cylinder rotate, overcoming inertia and the Faraday force that acts on the cylinder in direction opposite to its motion?

In Fig. 7.3, the symbols L_e and L_i denote the angular momentum of electrons and ions (i.e. the cylinder), the direction is obtained using the right-hand rule: the fingers point in the direction of rotation and the thumb points in the direction of the angular momentum vector.

To explain the Meissner effect, the London brothers wrote down Eqs. (6.4)–(6.6) from Chapter 6. Equation (6.6) was

$$\vec{\nabla} \times (\vec{v}_s(\vec{r},t) - \vec{v}_s(\vec{r},0)) = -\frac{e}{m_e c}(\vec{B}(\vec{r},t) - \vec{B}(\vec{r},0)). \tag{7.1}$$

The initial conditions are: $v_s(\vec{r},0) = 0$ and $\vec{B}(\vec{r},0) = \vec{B}$, uniform inside the cylinder. Therefore,

$$\vec{\nabla} \times \vec{v}_s(\vec{r},t) = -\frac{e}{m_e c}(\vec{B}(\vec{r},t) - \vec{B}(\vec{r},0)). \tag{7.2}$$

Clearly, the solution of this differential equation is

$$\vec{B}(\vec{r},t) = \vec{B}(\vec{r},0) \tag{7.3a}$$

$$\vec{v}_s(\vec{r},t) = 0. \tag{7.3b}$$

In other words, *nothing happens.* An electric current is not generated and the magnetic field in the interior doesn't change. This is what we said before, in a perfect conductor the magnetic field lines are 'frozen', they cannot change. However, this contradicts the experimental result of Meissner and Ochsenfeld.

For that reason, the London brothers *postulated* [2] that the correct equation, instead of Eq. (7.2), is

$$\vec{\nabla} \times \vec{v}_s(\vec{r},t) = -\frac{e}{m_e c}\vec{B}(\vec{r},t). \tag{7.4}$$

Equation (7.4) is the same as Eq. (6.7) and leads to Eq. (6.10), that says there is no magnetic field in the interior of the cylinder (except in a surface layer of thickness λ_L), that is, it describes the situation in the right panel of Fig. 7.3. Equation (7.4) is known as the

London equation. Or, more precisely, Eqs. (6.4) and (7.4) are known as the first and second London equations. As we saw in Chapter 6, Eq. (7.4) is derived from the laws of classical physics *assuming that initially there is no magnetic field in the interior of the cylinder.* The contribution of the London brothers was to postulate that Eq. (7.4) is valid in the superconducting state *independent of the initial conditions*, that is, it is also valid when initially there is a nonzero magnetic field in the interior of the cylinder. It doesn't look like a very important contribution? It is, especially because one of the brothers, Fritz, understood many of the fundamental consequences of this postulate in subsequent work that he performed.

It is frequently said that the second London Eq. (7.4) explains the Meissner effect. That is not so. Equation (7.4) describes a situation where there cannot be magnetic field in the interior of the superconductor. But that is not the Meissner effect. The Meissner effect is the *process* that leads from the initial state on the left panel of Fig. 7.3 to the right panel. London's treatment explains *nothing* about this process. On the contrary, according to Eq. (7.2), London's treatment predicts that the Meissner effect does NOT take place. Reality says it does. Neither did the London brothers formulate nor answer the question of how momentum is conserved when the Meissner current is generated. Neither did anybody else in later years.

The Meissner effect, the expulsion of magnetic field, occurs *in the process where the metal goes from being normal to being superconducting.* The presence of the magnetic field causes the current I to be generated during this process. If there is no magnetic field when the metal becomes superconducting, no current is generated. Or, perhaps one could imagine, currents in different directions are generated that in total cancel? BCS doesn't predict that, but we will talk about that possibility later on.

The correct explanation of the Meissner effect has to explain why the Meissner effect occurs when the laws of classical physics (1), (2), and (3) say that the Meissner effect should not occur. The experts in superconductivity tell us that quantum mechanics explains the Meissner effect. However, Bohr's correspondence principle mentioned earlier says that the laws of quantum mechanics have to coincide with the laws of classical physics when the system is macroscopic, which the system of Fig. 7.3 is. We will analyze in the following chapters what BCS theory says about the Meissner effect, and we will explain why only Heisenberg's holes can explain the Meissner effect.

References

[1] W. Meissner and R. Ochsenfeld, *Naturwissenschaften* **21**, 787 (1933).
[2] F. London and H. London, *Proc. Roy. Soc. A* **149**, 71 (1935).

Chapter 8

Essential points of BCS, and why it doesn't explain the Meissner effect

In short, *BCS theory proposes that superconductivity arises when electrons pair and condense in a quantum state with macroscopic phase coherence.* Let's see what this means.

First and foremost, electrons have to form pairs. That requires that there is an attractive interaction between electrons, but the Coulomb interaction between electrons is repulsive. How is it overcome?

BCS proposes [1] that the interaction of electrons with ionic vibrations in the solid (phonons) gives rise to a *retarded* interaction between electrons that is attractive. Qualitatively, the first electron goes by near a positive ion, it attracts it, and the ion approaches a little (see Fig. 4.1). A second electron that comes behind the first feels a stronger attractive interaction with this ion because it moved, and this corresponds to an effective attractive interaction between the first and the second electron. The interaction is *retarded* because there is a time interval between the times when the first and second electron are in the vicinity of the same ion, during which the ion moves.

The idea that superconductivity is due to interaction between electrons and phonons, that is lattice vibrations, was introduced by Herbert Fröhlich in 1950. We will discuss that part of the story, which is very important, later on. We will not enter into the details of this interaction here. Mathematically the treatment is not simple and, most importantly, it is difficult to ascertain exactly how this attractive interaction, denoted by the symbol λ, competes with the repulsive Coulomb interaction denoted by the symbol μ^*, in particular which wins. The experts maintain that they know how to calculate λ from first principles, but in general admit they don't know how to calculate μ^*, and simply postulate that it is sufficiently small that $\lambda - \mu^*$ is positive for superconducting materials. With this condition, they deduce that the system enters into the superconducting state at low temperatures. But there is in general no way to predict for which materials $\lambda - \mu^*$ is positive and for which ones it is negative, consequently they cannot predict without knowing the answer in advance, which materials can become superconducting and which cannot.

Why is it necessary that electrons pair? Because electrons are fermions, with fractional 'spin' (intrinsic angular momentum) 1/2, that obey what is called the Pauli exclusion principle, which prohibits that they form a collective state like superconductivity is. When they pair, the spin of the pair becomes integer, entities with integer spins are called bosons, and it is known since 1924 that bosons can condense in a macroscopically coherent state as the superconductor is, in a process known as 'Bose–Einstein condensation'. The same process is believed to account for liquid helium becoming a superfluid at low temperatures. We will consider liquid helium in Chapter 20.

Bardeen, Cooper, and Schrieffer (BCS) showed in 1957 [1] how, using pairs of fermions, a macroscopic quantum state can be formed that has properties that appear to describe a superconductor. The BCS wavefunction is of the form

$$|\Psi>=\prod_k (u_k + v_k c^\dagger_{k\uparrow} c^\dagger_{-k\downarrow})|0> . \tag{8.1}$$

The index k indicates the momentum of each electron, the operator $c^\dagger_{k\uparrow}$ creates an electron with spin up and momentum k, the operator $c^\dagger_{-k\downarrow}$ creates an electron with spin down and momentum $-k$. The combination $c^\dagger_{k\uparrow} c^\dagger_{-k\downarrow}$ creates what is known as a Cooper pair, because Cooper, one of the authors of BCS, proposed such pairs one year earlier, in 1956. The BCS formalism tells us how to calculate the amplitudes v_k, u_k given certain characteristics of the electronic states in the solid, the interactions λ and μ^*, etc.

The quantum state described by the BCS wavefunction, Eq. (8.1), has lower energy than the normal state of electrons in a metal provided that $\lambda - \mu^* > 0$. The formalism gives that the superconducting state Eq. (8.1) can only exist at low temperatures, below a critical temperature T_c that depends on $\lambda - \mu^*$, on other electronic properties of the solid, and, very importantly, on the frequency of vibration of the ions, called ω_D, the Debye frequency. The expression for T_c is

$$T_c = C\theta_D e^{-1/(\lambda-\mu^*)} \tag{8.2a}$$

where C is an adimensional numerical constant of order 1 and θ_D is the Debye temperature, proportional to ω_D. The Debye frequency is inversely proportional to the square root of the mass of the ion, M. Hence, an equivalent form of Eq. (8.2a) is

$$T_c = CM^{-1/2}e^{-1/(\lambda-\mu^*)} \tag{8.2b}$$

with another constant C. We will talk about this formula again in Chapter 12, in relation with what is called the 'isotope effect', the variation of the critical temperature with different isotopes of an element.

Below this critical temperature, the system described by this theory has very interesting properties: electrons are paired, forming Cooper pairs, and there is a finite energy gap, called Δ, that has to be paid to break a pair. Because of this, Cooper pairs conduct electricity without resistance: in the presence of electric current, they acquire the form $c^\dagger_{k+q\uparrow}c^\dagger_{-k+q\downarrow}$, that is they have a net moment $2q$, so they can conduct electricity until the current reaches a critical value where the pairs and the superconductivity are destroyed. For lower values of the current, the system conducts electricity without resistance, like superconductors do. BCS also calculated various properties of the system described by their wavefunction that can be measured experimentally, for example, the specific heat and the density of electronic states, and found results coincident with experiments on superconductors.

Finally, the BCS wavefunction Eq. (8.1) has macroscopic phase coherence: the complex number v_k/u_k has a phase φ that is independent of k, that is, it is common to the entire system. This is an essential characteristic of superconductors, that was experimentally confirmed with the observation of the so-called 'Josephson effect', predicted by British physicist Brian Josephson in 1962 and verified experimentally shortly thereafter. The Josephson effect shows that a superconductor behaves in essence like a giant quantum particle, with a macroscopic wavefunction $\Psi(r)$ with amplitude and phase, that is analogous to the microscopic wavefunction describing a single electron.

However, what BCS doesn't explain is the Meissner effect.

How can this be? the reader may say. BCS was formulated 60 years ago, everybody agrees that it describes the physics of superconducting materials, and one of the fundamental properties of superconductors is that they expel magnetic fields, i.e. the Meissner effect. This doesn't make sense.

Exactly. The problem is that BCS said in their 1957 paper that their formalism explains the Meissner effect, and everybody believed it.

Well, not quite. At the beginning, many physicists doubted that BCS explained the Meissner effect, for a technical reason: the proof that BCS had presented in their article [1] to prove that the theory predicts the Meissner effect did not satisfy 'gauge invariance', a necessary condition for a theory to be consistent. During a couple of years after 1957 several physicists, for example Anderson, one of the 'villains' of Chapter 2, made complicated calculations that finally showed that it was possible to extend the BCS formalism so that it would satisfy gauge invariance [2]. From there on, the scientific community took as an article of faith that BCS theory explains the Meissner effect.

It reminds me of the plot of several of Agatha Christie's detective novels: the true criminal plants clues against him/herself that are discovered early on, making him/her the first suspect. Later it is found that this evidence is false, and from there on nobody doubts the innocence of this person, the discarded initial suspicion 'immunizes' him/her. Until much later, at the end of the story, it is revealed that that person was indeed the criminal, with different evidence than the initial false evidence that was discarded.

Similarly here. The initial doubts were resolved, and this made BCS' explanation of the Meissner effect immune to being questioned. When I say that BCS theory does not explain the Meissner effect, the response is: this was already considered earlier and it was proven that it does.

But it is not so, for the following reason. The calculation that BCS does for the Meissner effect [1], just like the calculations that Anderson [2] and others do, start from the assumption that the system is described by the BCS wavefunction, Eq. (8.1), then a small magnetic field is applied, and the system responds modifying the wavefunction Eq. (8.1) with an additional term that generates the Meissner current near the surface that nullifies the magnetic field in the interior.

That is *not* explaining the Meissner effect, because the starting point is the BCS wavefunction Eq. (8.1). That wavefunction *cannot describe a system with a magnetic field in its interior.*

The initial state in Meissner's experiment is the *normal state* with a magnetic field in the interior, that is the left panels in Figs. 7.2 and 7.3, not the BCS state. The system cannot first adopt the BCS state and then expel the magnetic field, because the BCS state with phase coherence cannot exist in the presence of an interior magnetic field. To explain the Meissner effect, one has to explain how the system evolves dynamically from the initial normal state to the final state, a state with surface current and without interior magnetic field, described by the right panels of Figs. 7.2 and 7.3. The BCS formalism does not consider that question.

I say that BCS theory doesn't have the physical elements necessary to describe the dynamics of the Meissner effect, in particular to explain how the system manages to in appearance violate the laws of classical physics previously discussed. There is nothing in the physics of electron pairing, electron–phonon interaction, energy gap, macroscopic phase coherence, etc, that would explain the processes shown in Figs. 7.2 and 7.3, the expulsion of magnetic field when a normal metal in the presence of a magnetic field enters the superconducting state.

References

[1] J. Bardeen, L. N. Cooper, and J.R. Schrieffer, Theory of superconductivity, *Phys. Rev.* **108**, 1175 (1957).

[2] P. W. Anderson, Coherent excited states in the theory of superconductivity: Gauge invariance and the Meissner effect, *Phys. Rev.* **110**, 827 (1958).

Chapter 9

The Meissner effect in more detail

Let us consider in more detail the process of expulsion of a magnetic field B from the interior of a metal that is cooled and enters the superconducting state. Figure 9.1 shows three conceivable routes.

In the top route (a), a surface current $I(t)$ is generated that circulates in a layer of thickness λ_L (London penetration depth) next to the surface, it gradually increases its magnitude generating an interior magnetic field that gradually compensates the external applied field, until it reaches the magnitude needed to completely cancel the magnetic field in the interior, right panel.

It turns out that route (a), that appears to be the simplest, is not feasible, for the following reason: the energy of the superconducting state is lower than that of the normal state, and part of that energy difference (called 'condensation energy') has to be used to give energy to the growing current $I(t)$. However, the system cannot enter the superconducting state until the magnetic field is completely eliminated in the interior, hence there is no energy available to generate the surface current in route (a). In this, BCS and I agree.

Route (b) is energetically possible. There, superconducting domains nucleate where the current circulates on their surface (black curved arrows), canceling the magnetic field in the interior of each domain. The energy to generate these currents comes from the condensation energy inside each domain. The domains gradually expand (red arrows) and eventually coalesce to cancel the magnetic field in the entire interior, as shown on the right panel.

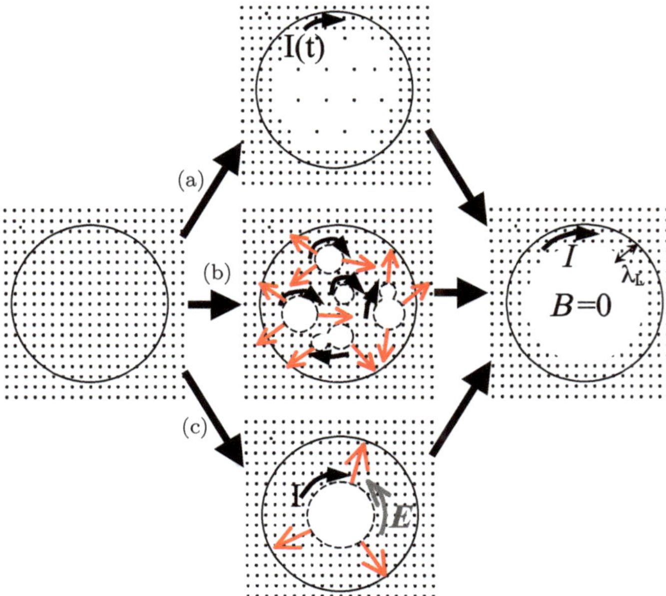

Fig. 9.1 Cylinder seen from the top. Initially (left panel) the cylinder is in the normal state and a magnetic field exists throughout its interior pointing perpendicular to the paper, denoted by the dots. The density of dots denotes the intensity of the magnetic field. The figure shows three conceivable routes to expel the magnetic field and reach the right panel where there is no magnetic field in the interior and current circulates near the surface. The red arrows indicate the expansion of the superconducting domains. The text explains the processes in detail.

Route (c) is a simplification of route (b), where there is a single central domain that expands gradually until the right panel is reached. Experimentally it can be realized by putting a radial temperature gradient, where the interior is a little colder. What is important is that the physics is essentially the same in routes (b) and (c), so we can concentrate on route (c), which is simpler, to understand what happens.

What happens is that in the process of the metal becoming superconducting, the superconducting region gradually *expands*. The essence of the process is *expansion* of the superconducting region or regions. The *expansion* produces the *expulsion* of the magnetic field.

BCS would agree that the magnetic field is expelled through expansion of one or many domains. But it has nothing to say about:

(1) What is the force that makes the domain expand?
(2) What is the force that propels the current around the contour of each domain? The energy and angular momentum of the current grow as the domain grows.
(3) How is the momentum of the currents around the domains compensated, to satisfy momentum conservation?
(4) How is phase coherence in the interior of each domain established?
(5) How is Faraday's law overruled? Around each growing domain, a Faraday electric field E is generated in counterclockwise direction (indicated in the lower central panel of Fig. 9.1) that opposes the generation of clockwise current that develops to nullify the field B in the interior of the domain, nevertheless the current is generated.

To all these questions, physicists that believe in BCS (i.e. all physicists) have a single answer: *'The only way to enter the superconducting state is expelling magnetic field, since the superconducting state cannot exist in the presence of magnetic field. Energetically, it is favorable for the domains to enter the superconducting state, paying the energetic price to generate the current. So it will happen, the details don't matter. Physical systems always have the tendency to evolve to states of lower energy. In fact, one can define the 'generalized force' that propels the current as the derivative of the energy with respect to a generalized coordinate.'*

Faraday's law, the fact that during the process an electric field is generated that opposes the process and one has to explain how it is overcome, is not even considered. They are not even aware that Faraday's field exists, since it is not there in the initial and final states, and all they care about is initial and final states.

That explanation doesn't explain anything. It does not answer the questions posed above. And in addition, it is false. Examples where physical systems DON'T evolve to states of lower energy are the rule, not the exception, in nature. For example, if the reader is on the second floor of a building, would s/he 'spontaneously' evolve to the ground floor where s/he would have less gravitational energy? There has to be a *way*, that respects the laws of physics, to connect the initial and final states. A way that explains the *dynamics* of the process. BCS doesn't provide it for the Meissner effect. In the example of the building, the way would include a staircase or an elevator. If there is neither, the reader could either stay on the second floor forever, or alternatively jump off the balcony to reach the ground floor, but the end result would be different than the one observed, it would include some broken bones. The correct way has to connect the observed initial and final states.

How do I know that BCS says what I said above? Not because it is in the scientific literature, nobody poses these questions. I know it because in innumerable referee reports on my papers where I have posed these questions I was told that there is no need to pose them because BCS answers them, in the way explained above. And, because in innumerable discussions I have had in person and through email with superconductivity experts such as those listed in Chapter 2 and many others, that is the clarification I get from them: none.

Chapter 10

The key to the Meissner effect: charge expulsion

Let us try to understand the physics of the Meissner effect qualitatively. As we saw in the right panel of Fig. 7.3, electrons near the surface of the cylinder acquire a velocity in counterclockwise direction when the system goes from the normal to the superconducting state in the presence of a magnetic field pointing upward. BCS theory does not explain the dynamics of the process that appears to violate the laws of classical physics. Can we imagine some process that explains this in a simple way?

The answer is yes as shown in Fig. 10.1.

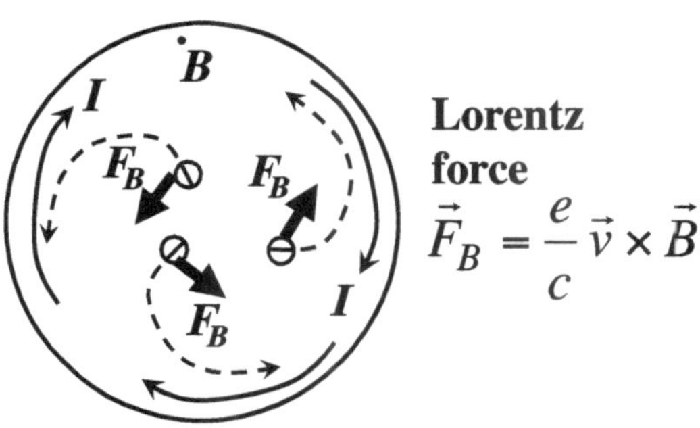

Fig. 10.1 The figure shows a metal cylinder seen from the top, that initially has a uniform magnetic field B throughout its interior pointing out of the paper. If electrons are expelled from the interior radially outward, the Lorentz force deviates them in counterclockwise direction along the trajectory indicated by the dashed lines, resulting in a clockwise current I near the surface.

If there is a radial flux of electrons from the interior of the cylinder towards the surface, and a uniform magnetic field in the interior parallel to the cylinder axis pointing out of the paper, the Lorentz force acting on the electrons is in counterclockwise direction. When the electrons reach the surface with the acquired azimuthal velocity, they give rise to a clockwise surface current, which generates a magnetic field in the interior pointing into the paper, that is in opposite direction to the applied field B. If that current is sufficiently large, it will completely suppress the magnetic field in the interior of the cylinder, which means that the magnetic field was expelled from the interior. The same occurs in the domains of Fig. 9.1. The *expansion* of the domains, i.e. the outward motion of the border of each contour, has to be associated with an *outward flux* of electrons.

That is the physical origin of the Meissner effect. I claim that *the only way* to expel a magnetic field from the interior of a metal, or of a region of a metal, is to expel electric charge. As charge is expelled, we may think that it 'drags' the magnetic field lines with it, since the interior field decreases as explained above.

The electric charge that is expelled could also be positive, the Lorentz force would be in opposite azimuthal direction and the electric current generated would be in the same direction as shown in Fig. 10.1, generating a magnetic field that suppresses the interior field. So one could think that if 'holes' are expelled rather than electrons, it would be an alternative way to explain the Meissner effect. That is not so, the primary mechanism is expulsion of negative charge, as I will explain later. But in fact the process is slightly more complicated than what I described above, it involves radial motion of *both* electrons and holes.

This leads us to conclude that in the process of becoming superconducting, the metal expels negative charge from the interior toward the surface. That is the *essence* of the Meissner effect. It is a simple and intuitive explanation, and for that reason it is very likely correct. I don't believe there is any other possible explanation of the *dynamics* of the process, certainly nobody has proposed an alternative.

In fact, it is not difficult to prove mathematically that if there is no charge expulsion there cannot be a Meissner effect. The equation of motion for an electron of charge e, mass m_e, in the presence of electric and magnetic fields is more complicated than Eq. (6.4), that was an approximation. The correct equation is

$$\frac{d\vec{v}}{dt} = \frac{e}{m_e}\vec{E} + \frac{e}{m_e c}\vec{v} \times \vec{B}. \tag{10.1}$$

Using the mathematical relation between partial and total time derivatives

$$\frac{d\vec{v}}{dt} = \frac{\partial \vec{v}}{\partial t} + (\vec{v} \cdot \vec{\nabla})\vec{v} = \frac{\partial \vec{v}}{\partial t} + \vec{\nabla}(\frac{\vec{v}^2}{2}) - \vec{v} \times (\vec{\nabla} \times \vec{v}) \tag{10.2}$$

and Faraday's law $\vec{\nabla} \times \vec{E} = -(1/c)\partial \vec{B}/\partial t$, it follows from Eq. (10.1) that

$$\frac{\partial \vec{w}}{\partial t} = \vec{\nabla} \times (\vec{w} \times \vec{v}) \tag{10.3}$$

for the 'generalized vorticity'

$$\vec{w} = \vec{\nabla} \times \vec{v} + \frac{e}{m_e c}\vec{B}. \tag{10.4}$$

In the initial normal state, when $t = 0$,

$$\vec{w}(\vec{r}, t = 0) = \frac{e}{m_e c}\vec{B}(t = 0) \equiv \vec{w}_0 \tag{10.5}$$

independent of position $\vec{r}$, since $\vec{\nabla} \times \vec{v} = 0$ because there is no current. In the superconducting state, the carrier velocity satisfies the London equation (7.4)

$$\vec{\nabla} \times \vec{v} = -\frac{e}{m_e c}\vec{B}. \tag{10.6}$$

Therefore, according to Eq. (10.4)

$$\vec{w}(\vec{r}, t = \infty) = 0 \tag{10.7}$$

for all positions inside the superconductor. Hence, the Meissner effect is the time evolution of the quantity $\vec{w}(\vec{r}, t)$ from its initial nonzero value given by Eq. (10.5) to its final value zero everywhere in the interior of the superconductor.

Now with cylindrical symmetry around the z axis we have

$$\vec{w}(\vec{r},t) = w(r,t)\hat{z} \tag{10.8}$$

and the time evolution of w, Eq. (10.3), takes the form

$$\frac{\partial w}{\partial t} = -\frac{1}{r}\frac{\partial}{\partial r}(rwv_r) \tag{10.9}$$

with r the radius in cylindrical coordinates, and v_r the component of the electrons' velocity *in radial direction.*

We used Faraday's law to derive Eq. (10.9). Equation (10.9) tells us that $w(r,t)$ can only change in time if there is radial flux of charge, $v_r(r,t) \neq 0$ at that point r at that instant. Additionally, for w to be able to reach its final state $w = 0$, corresponding to the Meissner state, requires that $v_r > 0$, that is radial flux of charge from inside out. In other words, Faraday's law forbids expulsion of magnetic field unless there is an accompanying expulsion of charge.

We can illustrate this further with the following analogy. As was shown in Figs. 7.1 and 7.2, in the Meissner effect, a magnet is lifted by the metal becoming superconducting against the force of gravity in a process that seems to contradict the predictions of Faraday's law. According to the conventional theory of superconductivity, this is caused by a mysterious force that does not exist in the normal metallic state. Instead we may ask: would it be possible *for a normal metal* to lift a magnet without contradicting Faraday's law? The answer is yes, as illustrated in Fig. 10.2.

The upper panel of Fig. 10.2 shows how the Meissner effect lifts a magnet, and the lower panel shows how a normal metal can lift a magnet. In the lower panel, a copper tube is placed around a magnet, the tube is slightly wider than the diameter of the magnet so it doesn't touch it. Yet as we raise the tube rapidly the magnet moves along with it, so it is lifted. This happens precisely because of Faraday's law rather than in contradiction with Faraday's law.

Why is the magnet lifted by the metal tube? We can assume that the magnet never touches the inner wall of the tube. Nevertheless, as the tube moves up, magnetic field lines from the magnet start to "cut through" the metal, and this generates eddy currents in the metal

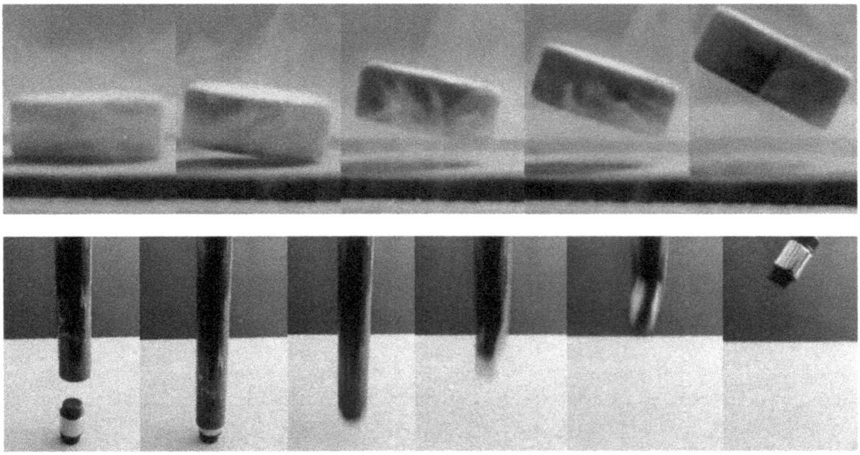

Fig. 10.2 Lifting of a magnet by a superconductor and by a normal metal. In the upper panel, a magnet rests on a normal metal disk initially (upper left panel) and is *lifted* by the Meissner effect when the metal is cooled and becomes superconducting. In the lower panel, a metal tube is put around a magnet and the tube is rapidly raised. The magnet is *lifted* by the motion of the electric charges in the walls of the metal tube. We argue that the explanation is the same for the upper and lower panels, namely Faraday's law.

tube that create a magnetic field that interacts with the magnetic field of the magnet so as to oppose motion of the magnet relative to the metal tube. The physical principle governing this is precisely Faraday's law of induction. The better conducting the metal is, the more it opposes magnetic field flux changes (Lenz's law).

In the case of the magnet and the metal tube, Fig. 10.2 lower panel, we understand the lifting of the magnet based solely on the laws of classical electromagnetism, Maxwell's equations, which include Faraday's law. The charge expulsion discussed earlier in this chapter is realized in this analogy by the upward motion of the metal tube. If in the upper panel of Fig. 10.2 we have charge expulsion, we have a common explanation for the phenomena shown in the lower and upper panels of Fig. 10.2: the lifting of the magnet is explained by Faraday's law in both cases. Instead, according to the accepted theory of superconductivity since 1957 (BCS theory), the lifting of the magnet in the upper panel of Fig. 10.2 (Meissner effect) is not explained by the same physics as the lifting of the magnet in the lower panel of Fig. 10.2, but by completely different physics.

However, according to a well established rule of scientific reasoning, formulated by Isaac Newton in 1686, '*Numquam ponenda est pluralitas sine necessitate.*' That is, '*To the same natural effects* (in our case, lifting of the magnet in Fig. 10.2 upper and lower panels) *we must, as far as possible, assign the same causes.*'

We can apply the same rule to connect magnetohydrodynamics and the Meissner effect. The theorem of Hannes Alfven, mentioned in Chapter 2, states that in a perfectly conducting fluid the magnetic field lines are frozen into the fluid. In books that deal with the physics of plasmas, they show diagrams like the left panel of Fig. 10.3 to illustrate Alfven's theorem [1]. A jet of fluid moves from left to right, and causes the magnetic field lines to bend as the figure shows. *The same natural effect*, bending of magnetic field lines, occurs in the right panel. It is natural to assume that the same physics of the left panel

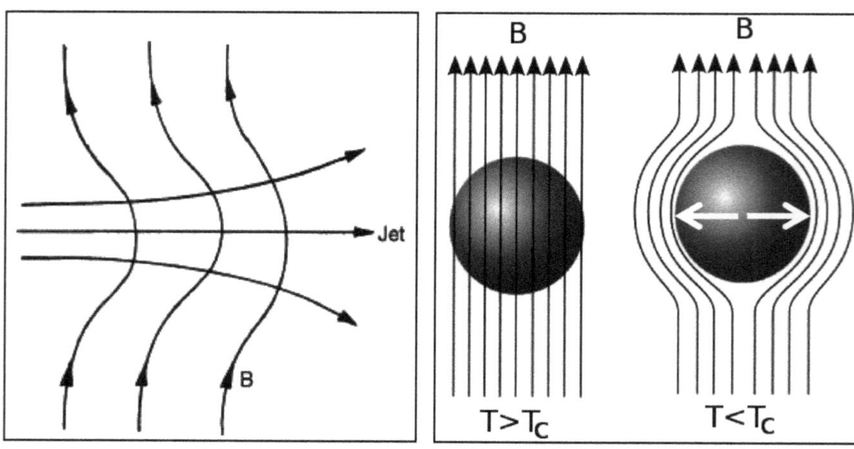

(a) (b)

Fig. 10.3 Analogy between the Meissner effect and processes that occur in conducting fluids called plasmas. According to Alfven's theorem illustrated on the left panel, a jet of plasma flowing from left to right gives rise to the curvature in the magnetic field lines B shown [1]. Similarly, the right panel shows how the expulsion of magnetic field from the interior of a metal that is cooled and enters the superconducting state, i.e. the Meissner effect, is explained by expulsion of charge, denoted by the white arrows.

is the *cause* of the physics of the right panel, expulsion of magnetic field from the interior of a superconductor when the Meissner effect takes place. Later we will explain how this outflow of charge is compensated, and how angular momentum is conserved in this process.

BCS theory *does not* predict charge expulsion, it says that $v_r = 0$, and for that reason I say that BCS theory *cannot* explain the Meissner effect. For a metal described by the physics contained in BCS theory, the quantity w in Eq. (10.9) does not change with time, and therefore cannot expel magnetic fields, therefore does not explain the Meissner effect, therefore it does not explain the superconductivity of real materials, all of which exhibit the Meissner effect.

What do BCS believers (i.e. all physicists) say about this? That the treatment I made above is a classical treatment, that plasmas are classical systems, but superconductors are quantum systems. That quantum mechanics makes superconductors expel magnetic fields. However, they don't tell us how. I say that according to Bohr's correspondence principle, quantum mechanics has to agree with classical mechanics when the system is in the macroscopic domain, and Faraday's law has to be respected. Therefore, expulsion of magnetic field when a metal is cooled and enters the superconducting state in the presence of a magnetic field *requires* $v_r \neq 0$.

Reference

[1] P. A. Davidson, *An Introduction to Magnetohydrodynamics*, Cambridge University Press, Cambridge, 2001.

Part III

CONVENTIONAL THEORY AND THE NEW MATERIALS

Chapter 11

Theories of superconductivity before BCS, that explain more than BCS

It is interesting to travel back in time and study theories of superconductivity that were formulated before BCS. They didn't get very far, but they had very interesting physical ideas. Instead, since BCS physicists stopped doing physics and devoted themselves to doing math, and so they lost touch with reality. In developing the theory of hole superconductivity, I discovered that several things that I was understanding had been understood before, but were forgotten after BCS was developed.

It is not infrequent in the history of science that early ideas are discarded using arguments that later are found not to be valid. A well-known example is the heliocentric model of the solar system proposed by Aristarchus in the year 200 BC, later discarded in favor of Ptolemy's geocentric model that was accepted for the next 15 centuries, only to be replaced again by the heliocentric model of Copernican astronomy. Originally, the Aristarchus model was considered invalid because it was thought that motion of the earth should give rise to 'wind' in direction contrary to the motion, that was not observed, and because it was thought that one should observe resulting motion of distant stars, which was not observed. Both arguments were incorrect, the first because the atmosphere moves with the earth, the second because the distant stars turned out to be much more distant than what was thought at the time.

Another less well-known example is the planetary model of atoms, originally proposed by Jean Perrin and Hantaro Nagaoka in the years 1901–1904, then discarded in favor of the 'plum pudding' model of J. J. Thomson, a very prominent physicist awarded the 1906 Nobel prize for the discovery of the electron. The model of Perrin and Nagaoka was not accepted because electrons orbiting around a nucleus are accelerated, and according to Maxwell's theory of electromagnetism they should radiate and lose energy, which does not happen. The Thomson model resolved this 'problem'. Yet a few years later, the Thomson model was itself discarded, replaced by the planetary model of Ernest Rutherford and Niels Bohr. Bohr simply postulated, correctly, that Maxwell's equations did not apply equally at the atomic domain and the macroscopic domain.

In what follows, we discuss several aspects of these pre-BCS theories of superconductivity that were not incorporated into BCS theory, that reappear in some form in the theory of hole superconductivity.

11.1 The Coulomb interaction as responsible for superconductivity

In practically all pre-BCS theories of superconductivity, it was assumed that the Coulomb interaction between electrons plays a fundamental role. That was very reasonable for the following reason: Bloch's theory, developed in 1928 to explain properties of normal metals, prioritizes the electron–ion interaction, and ignores the electron–electron interaction. There is no theoretical justification for this, since both interactions give rise to energies of order e^2/r, with r a typical distance that is of the same order for electron–ion and electron–electron interactions. The justification for Bloch's hypothesis was simply that it gives rise to predictions *for the normal metal* that explain the observed behavior. Therefore, it is very natural to conclude that the big differences in the behavior of normal metals and superconductors originate in 'the elephant in the room', the large Coulomb interaction between electrons ignored by Bloch. Bloch himself clearly hypothesized this in his original papers, as did many other theorists at that early time. Quite on the contrary,

BCS theory continues to ignore the electron–electron interaction and emphasizes the electron–ion interaction even more, by proposing that the tiny (relatively speaking) energy resulting from the interaction of electrons with the small deviations of the ions from their equilibrium positions, i.e. the vibrations of the ions, is what causes superconductivity.

As the eminent experimental physicist K. Mendelssohn wrote in 1966, '*To the layman it may come as a disappointment that the explanation of such a striking phenomenon as superconductivity should, on the atomistic scale, have been revealed as nothing more exciting than a footling small interaction between electrons and lattice vibrations. This feeling was shared by many physicists who had hoped that superconductivity might reveal some new fundamental principle of nature*', undoubtedly including himself in that physicists' group.

Instead, within the theory of hole superconductivity, the Coulomb interaction between electrons plays a key role, and the interaction of electrons with ionic vibrations plays no role.

11.2 Circulating currents in the superconducting state

Many theorists before BCS proposed models with 'spontaneous currents' or 'circulating currents', to explain superconductivity. Among them Bloch [1], Frenkel [2], Landau [3], Smith [4], Born, Cheng [5], and Heisenberg [6].

In these models, circulating electric currents exist in the superconductor even in the absence of external fields. It was argued that one doesn't see a manifestation of these currents because there are 'domains' in the material with different orientation in the motion of these currents, so that no net magnetic field is generated. When an external field is applied, the domains change their orientation so as to give rise to the currents and magnetic moments that one observes. These current domains would be analogous to the magnetization domains that are known to exist in ferromagnetic materials.

Even before the Meissner effect was discovered, theorists such as Frenkel and Landau proposed such models. Frenkel [2] in particular pointed out that a consequence of this physics would be that the material would be highly diamagnetic, which is precisely what the Meissner effect shows. But neither Frenkel nor others explained the mechanism by which generation of these currents expels an applied magnetic field in the normal-superconductor transition.

This assumption, that there are currents in the superconductor in the absence of applied external fields, is very natural, since intuitively it makes sense that it is easier to generate a supercurrent by orienting domains of preexistent currents instead of generating it from zero. In a way, it is what occurs in atoms: the atomic currents preexist in the orbits of electrons in atoms, an applied magnetic field orients them, it doesn't generate them.

In BCS theory, there are no preexistent currents. That is, this concept that was shared by so many early theorists, is totally absent. Instead, in the theory of hole superconductivity, there are preexistent currents [7].

I will explain those currents within the theory of hole superconductivity in Chapter 19. They differ from the currents proposed by these early theorists in that they are what is called 'spin currents' rather than charge currents. In essence, electrons with intrinsic angular momentum (spin) pointing up circulate in one direction and electrons with spin down circulate in opposite direction. Due to this, it is not necessary to have domains since the charge current is zero and no magnetic moment is generated in the absence of external fields. The orientation of these spin currents is determined by the geometry of the superconducting body.

11.3 Mesoscopic orbits

In 1937, J. C. Slater published a very interesting paper [8] where he proposed that electrons in superconductors reside in orbits, and that "*to produce superconductivity the orbits must be of the order of magnitude of 137 atomic diameters*". Slater's paper points out that the magnetic susceptibility of an atom is proportional to the square of the radius of the electronic orbits, and calculates that to have perfect diamagnetism, the orbits have to have diameter $137 \times 2a_0$, where a_0 is the Bohr radius, 0.529 Å.

Where does the number 137 come from? 1/137 is approximately the magnitude of the 'fine structure constant' α, defined by the formula

$$\alpha = \frac{e^2}{\hbar c} \sim \frac{1}{137} \tag{11.1}$$

where $\hbar$ is one of the fundamental constants in physics, Planck's constant. α plays an important role in many areas of physics, in particular, in quantum electrodynamics. Let's understand Slater's argument.

The Bohr radius is $a_0 = \hbar^2/(m_e e^2)$. Assume we have atoms in a cubic lattice with lattice constant (distance between nearest neighbor atoms) $2a_0$, and one electron per atom. The density of electrons is then

$$n_s = \frac{1}{(2a_0)^3}. \tag{11.2}$$

Using Eq. (6.10b) for the London penetration depth λ_L, we obtain

$$\frac{1}{\lambda_L^2} = \frac{4\pi n_s e^2}{m_e c^2} = \frac{\pi}{a_0^2}\left(\frac{e}{\hbar c}\right)^2 = \frac{\pi}{(137 a_0)^2} \tag{11.3}$$

therefore Slater's orbits are of diameter of the order $137 \times 2a_0 = 2\sqrt{\pi}\lambda_L$.

Now the diamagnetic susceptibility, also called 'Larmor susceptibility' for a solid of atomic density n_s is, as we learn in solid state physics books [9]

$$\chi = -\frac{n_s e^2}{4 m_e c^2} < r^2 > \tag{11.4}$$

where $< r^2 >$ is the average of the square of the radius of atomic orbits. The quantity χ is dimensionless, in a typical solid with $< r^2 > \sim 1\,\text{Å}^2$ its magnitude is of order 10^{-6}. In a superconductor, diamagnetism is perfect, that corresponds to $\chi = -1/(4\pi)$, i.e. the diamagnetism is 100,000 times stronger than in a normal metal. Perfect diamagnetism means that when a magnetic field is applied to a body it responds by generating an equal and opposite magnetic field in its interior, such that the magnetic field in the interior remains zero.

Using expression (11.3) for the London penetration depth, we obtain from Eq. (11.4) assuming the atomic radius is $r \sim 2\lambda_L$

$$\chi = -\frac{n_s e^2}{4m_e c^2} < (2\lambda_L)^2 >= -\frac{1}{4\pi}. \tag{11.5}$$

In other words, if the orbits have radius $2\lambda_L$, the solid is perfectly diamagnetic, it is a superconductor. Slater's orbits have radius $\sqrt{\pi}\lambda_L = 1.77\lambda_L$. Almost the same. These orbits are mesoscopic, of radius of order hundreds of Angstrom. Recall that 1 Angstrom is one tenth of a nanometer.

In BCS theory, mesoscopic orbits don't play any role. Instead, as we will see in subsequent chapters, in the theory of hole superconductivity, electronic orbits of radius $2\lambda_L$ play a fundamental role. I came to this conclusion without knowing about the existence of this paper by Slater, that has only 25 citations (13 of which are from my papers).

11.4 Superconductors as giant atoms

In several early papers on superconductivity, it was talked about superconductors as if they were 'giant atoms'.

For example, the first paper by the London brothers in 1935 [10] says '*Wenn wir uns den Supraleiter als ein grosses diamagnetisches Atom vorstellen*', i.e. '*When we imagine the superconductor as a giant diamagnetic atom*' as justification for the second London equation, since that equation describes the behavior of a diamagnetic atom upon application of a magnetic field.

In a paper in 1937, Fritz London [11] says '*The properties (1) and (2) characterize the electromagnetic behavior of the superconductor as being the same as that of a single big diamagnetic atom*'. The properties (1) and (2) that he is referring to are that the energy of the ground state is separated from the energy of excited states by an energy gap, and (2) that the quantum wave function that describes the superconductor is not modified to first order when a magnetic field is applied, in other words it is 'rigid'.

In an extensive review article on superconductivity in 1935, their authors Smith and Wilhelm [12] explain '*The electronic structure is to be considered as made up of aggregates which behave like diamagnetic atoms of sufficient size to make the body as a whole a perfect diamagnetic; on account of the size of these aggregates classical theory should give a fair first approximation to their properties and behavior*'.

In a 1953 article 'Der gegenwärtige Stand der Theorie der Supraleitung' ('The current state of the theory of superconductivity'), his author W. L. Ginsburg (famous for Ginsburg–Landay theory) says [13]: *Ein riesiges Atom von makroskopischen Abmessungen und mit einer entsprechend hohen Elektronenzahl würde sich in einem Magnetfeld wie ein Supraleiter verhalten ... Das Wesen der diamagnetischen Hypothese der Supraleitung besteht in diesen Überlegungen, nämlich in der Analogie zwischen einem Supraleiter und einem makroskopischen diamagnetischen Atom*', meaning: '*A giant atom of macroscopic dimensions and with a correspondingly large number of electrons would behave like a superconductor in a magnetic field ... The essence of the diamagnetic hypothesis of superconductivity is based on this reasoning, that is the analogy between a superconductor and a macroscopic diamagnetic atom*'.

After the advent of BCS theory in 1957, there was no longer talk of superconductor as macroscopic atoms. However, I believe this intuition of these early authors had a deep content of truth, even deeper than what those authors had understood. As we will see in Chapter 15, the theory of hole superconductivity predicts that superconductors are 'giant atoms' not only because of their magnetic properties that suggested this analogy to these early authors, but also because of their electrical properties, with respect to the spatial distribution of electric charge.

11.5 Charge expulsion

In the year 1940, the Austrian physicist Karl Michael Koch published a paper titled 'Versuch einer elektronenphysikalischen Deutung des Meissner-Ochsenfeld-Effektes' (An attempt at an electrophysical interpretation of the Meissner–Ochsenfeld Effect'. In it he pointed out that in the metal-superconductor transition there would be a temperature difference between the body becoming superconducting and its environment, which would give rise to a heat flow from the interior to the exterior of the body. Generally in metals heat flux is accompanied by an electric current, this is called the thermoelectric effect. Koch pointed out that during the transition this heat flow would give rise to an electric current flowing from the interior of the body toward the surface, and in the presence of a magnetic field the Lorentz force acting on this current would produce a circular current near the surface that would cancel the magnetic field in the interior.

This is practically the same argument I exposed in Chapter 10 to explain the Meissner effect. I didn't copy it from Koch, I proposed that in the transition from metal to superconductor there is expulsion of negative electric charge toward the surface of the body in the year 2001 [15] for reasons I will explain later, that have nothing to do with the thermoelectric effect, and in 2003 I proposed that this could be the explanation of the Meissner effect [16]. I didn't know about Koch's paper until 2009 when I found it by chance thanks to Google. I find it notable that Koch and I came to the same conclusion, that radial motion of charge explains the Meissner effect, through very different routes.

Nobody paid attention to this paper by Koch. In the 80 years since it was published to the present it was cited a total of only six times, of which two are self-citations and two are from papers I wrote in 2009 and 2013. Even though it describes for the first time in the scientific literature the basic principle that explains the Meissner effect, the most fundamental property of superconductors. In contrast, the BCS article that *does not* explain the Meissner effect has 7959 citations. That is how bibliometrics sometimes *doesn't* work.

11.6 Other examples

Finally I would like to mention some other examples of ideas or concepts before BCS that were not incorporated into BCS but I find that they have some or a lot of truth.

As I mentioned before, the idea that hole carriers play an important role in superconductivity was initially proposed by the Russian experimental physicists Kikoin and Lazarew [17] in the year 1932, and later was reiterated and emphasized by the Russian physicist Chapnik in 1962 and subsequent years. In addition, the physicist Papapetrou in 1934 motivated by Kikoin and Lasarew's article formulated a theory based on holes [18]. Physicists Born and Cheng in 1948 [5] also used in their theory that the charge carriers were holes, apparently they reached this conclusion independently of the work of Kikoin and Lasarew since they don't cite it. In BCS theory, holes don't play any special role, on the contrary, the theory has 'electron–hole symmetry'.

In the preface to his book on superconductivity in 1950 [19], Fritz London says "According to quantum theory the most stable state of any system is *not* a state of *static equilibrium* in the configuration of lowest potential energy. It is rather a kind of *kinetic equilibrium* for the so-called zero point motion, which may roughly be characterized as defined by the minimum average total (potential + kinetic) energy". And he adds "It is not necessarily a configuration close to the minimum of the potential energy (lattice order) which is the most advantageous one for the energy balance, since by virtue of the uncertainty relation the kinetic energy also comes into play. If the resultant forces are *sufficiently weak* and act between *sufficiently light* particles, then the structure possessing the smallest total energy would be characterized by a good economy of the kinetic energy ... It would be the outcome of a quantum mechanism of macroscopic scale."

These are profound concepts but they are not specific, and London did not elaborate on them in his book. In BCS theory, they are not taken into account, in particular, the state of minimum energy is *not* characterized by an economy of kinetic energy, on the contrary, the kinetic energy in the superconducting state is higher than in the normal state according to BCS. There is also no zero-point energy in BCS. In contrast, as we will see later, in the theory of hole superconductivity kinetic energy plays a central role and is economized in the superconducting state, just as London expected. In addition, there is macroscopic zero-point motion, as London had intuited.

With respect to the origin of the force that propels the current in the Meissner effect, I searched exhaustively in the early literature for what had been said on it. Except for Koch's work I found no other article that offered an answer, and what is more interesting, practically no article that asked the question, except one: in 1935, H. London wrote a very interesting article where he discusses equilibrium between normal and superconducting phases in a magnetic field [20], and comments "*The generation of current in the part which becomes supraconductive takes place without any assistance of an electric field and is only due to forces which come from the decrease of the free energy caused by the phase-transformation*". H. London does not specify the nature of these forces, obviously because he didn't know. But at least he formulated what is a key question, something that BCS doesn't do. I mention in passing that this article by Heinz London, that I consider to be of fundamental importance, has relatively few citations, 62 total in the 83 years since it was published, five of which are from his or his brother's papers and nine from my papers.

In conclusion, it is important to point out that many of the physicists mentioned in this chapter are extremely famous, but not for the papers cited in this chapter, that have very few citations. Felix Bloch, Werner Heisenberg, Lev Landau, and Max Born received the Nobel prize in physics in 1952, 1932, 1962, and 1954, respectively, for other very important works, recognized as such, that they performed in condensed matter physics or other areas of physics. About John Slater it is said that he 'almost' received the Nobel prize in 1977 for his extensive and important contributions to the electronic theory of atoms, molecules, and solids (but Slater passed away that year, and instead the prize was awarded to Phil Anderson, one of the 'villains' of Chapter 3). This suggests that the ideas about superconductivity formulated by these great physicists discussed in this chapter that were ignored were not necessarily nonsense.

I would also like to reiterate that the fact that the theory of hole superconductivity that I present in this book incorporates elements of these early theories of superconductivity doesn't mean that these theories influenced me in the development of the theory. On the contrary, I didn't know the papers by these scientists on these subjects as I was making progress in the development of the theory, and those papers are not discussed in the textbooks on superconductivity. I suggest that the fact that there is this relationship which is more than coincidental, it is an argument in favor of the validity of these ideas. In addition, the theory of hole superconductivity unifies all these ideas and shows their relationship, something that certainly wasn't clear to these authors that originally formulated them, as evidenced by the fact that they don't cite each other.

References

[1] F. Bloch, unpublished, referred to in H. Bethe and A. Sommerfeld, *Handbuch der Physik* XXIV, p. 12 (1933).
[2] J. Frenkel, On a possible explanation of superconductivity, *Phys. Rev.* **43**, 907 (1933).
[3] L. Landau, *Physik. Zeits. d. Sowjetunion* **4**, 43 (1933).
[4] H. Grayson Smith, Superconductivity in the light of accepted principles, Univ. Toronto Studies, Low Temp, Series, No. 76, 1935.
[5] M. Born and K. C. Cheng, Theory of superconductivity, *Nature* **161**, 968 (1948); *Nature* **161**, 1017 (1948).
[6] W. Heisenberg, Zur theorie der supraleitung, *Z. Naturforsch.* **2a**, 185 (1947); *Z. Naturforsch.* **3a**, 65 (1948).
[7] J. E. Hirsch, Spin currents in superconductors, *Phys. Rev. B* **71**, 184521 (2005).
[8] J. C. Slater, The nature of the superconducting state. II, *Phys. Rev.* **52**, 214 (1937).
[9] N. W. Ashcroft and N. D. Mermin, Solid State Physics, Holt, Rinehart, and Winston, Philadelphia, 1976.
[10] F. London and H. London, Supraleitung und diamagnetismus, *Physica* **2**, 341 (1935).
[11] F. London, On the nature of the superconducting state, *Phys. Rev.* **51**, 678 (1937).
[12] H. Grayson Smith and J. O. Wilhelm, Superconductivity, *Rev. Mod. Phys.* **7**, 237 (1935).
[13] W. L. Ginsburg, Der gegenwärtige stand der theorie der supraleitung, *Fortschritte der Physik* **1**, 101 (1953).
[14] K. M. Koch, Versuch einer elektronenphysikalischen Deutung des Meissner-Ochsenfeld-Effektes, *Z. Phys.* **116**, 586 (1940).
[15] J. E. Hirsch, Consequences of charge imbalance in superconductors within the theory of hole superconductivity, *Phys. Lett. A* **281**, 44 (2001).

[16] J. E. Hirsch, The Lorentz force and superconductivity, *Phys. Lett. A* **315**, 474 (2003).
[17] K. Kikoin and B. Lasarew, Hall effect and superconductivity, *Nature* **129**, 578 (1932).
[18] A. Papapetrou, Bemerkungen zur Supraleitung, *Z. Phys.* **92**, 513 (1934).
[19] F. London, *Superfluids*, Vol. I, Dover, New York, 1961.
[20] H. London, Phase-equilibrium of supraconductors in a magnetic field, *Proc. Roy. Soc. A* **152**, 650 (1935).

Chapter 12

Phonons and the beginning of obscurantism

Before 1950, nobody expected that the ionic vibrations in a solid (called phonons) had anything to do with superconductivity. In 1950, the coincidence of two experiments and one theory caused this to change. Since then, practically nobody doubts that superconductivity in the majority of materials, the so-called 'conventional superconductors', originates in the electron–phonon interaction.

In May 1950, two articles appeared in the journal *Physical Review*, with different authors but very similar titles: "*Isotope Effect in the Superconductivity of Mercury*", by Emanuel Maxwell [1], and "*Superconductivity of Isotopes of Mercury*", by Reynolds *et al.* [2]. These articles reported experiments that had found variations in the critical temperature of different isotopes of mercury, with atomic weight varying between Hg^{199} and Hg^{203}. The critical temperature varied by approximately 0.9% (from 4.126 K to 4.161 K), with the highest temperature corresponding to the smaller mass, when the mass varied between 199 and 203, i.e. by 2%.

This variation is very small but wasn't expected, and it generated great interest in the scientific community. Why? To begin with, generally different isotopes of an element have the same electronic properties. Hence, it was expected that the behaviour of electrons in superconductors would not be affected by the ionic mass. Especially surprising was the fact that the critical temperature increased when the mass of the ions decreased, in other words, the material conducted better. It was thought that lighter ions would vibrate more, so if vibrations increased sufficiently so as to modify the electrical resistance, which originates in collisions of electrons with vibrating atoms, the resistance should increase, not decrease, when the ionic mass was reduced.

John Bardeen, a theoretical physicist, learned about these experiments through a phone call from Serin, one of the authors of Ref. [2], on 15 May 1950. Immediately, Bardeen thought that this might be related to a paper he had written nine years earlier [3], in which he had proposed that static distortions of the ionic lattice would give rise to 'cells' containing approximately 10^6 atoms; this would reduce the 'effective mass' of the electrons to a very small value and give rise to superconductivity. Bardeen rapidly adapted these arguments to a dynamic distortion, and argued that this could give rise to a variation of the critical temperature proportional to the ionic frequency variation, and this would explain what the experiments of Maxwell [1] and Reynolds *et al.* [2] had measured. The ionic vibration frequency is inversely proportional to a $M^{1/2}$, with M the ionic mass, therefore this predicts $(\Delta T_c/T_c)/(\Delta M/M) = 1/2$, close to what was measured (0.9/2). As soon as one week later, on May 22nd, Bardeen submitted a paper for publication [4], of less than a page with only six equations, proposing a "*theory of superconductivity which depends on interaction of the valence electrons with the zero-point vibrations of the crystal lattice*" as the explanation for the observed 'isotope effect'.

Why the big rush? Because in science priority is important, the first to propose an explanation of a phenomenon gets the laurels, *if* the explanation is correct. But unfortunately for Bardeen, this explanation generated in such a rushed fashion that appears to be *ad hoc*, was not considered to be correct by anybody. Nevertheless, from that moment on Bardeen was convinced that the interaction of electrons with ionic vibrations (electron–phonon interaction) was the key to understand superconductivity, and he worked intensively in the following years based on that idea. Why?

To understand this, let's meet Herbert Fröhlich, a German physicist settled in England, where he was professor at the University of Liverpool. At that time, he was spending a sabbatical semester at

Purdue University in the United States. Fröhlich had never before worked on superconductivity. The same May of 1950, he submitted a paper to Physical Review [5] of 12 pages, with 88 equations, 5 figures, and 3 tables, where he proposed that the electron–phonon interaction is responsible for superconductivity. Notably, Fröhlich's article *does not mention* the experiments by Maxwell [1] and Reynolds *et al.* [2] that had measured the isotope effect.

Three days after sending his paper to *Physical Review* for publication, Fröhlich sent a Letter to the Editor [6] of the english journal *Proceedings of the Physical Society of London*, stating that the article that he had sent to *Physical Review* three days earlier *predicted* the experimental results of Maxwell and Reynolds *et al.* that had just been published. And when Fröhlich's article was published in *Physical Review* in September 1950 [5], it included a "note added in proof" saying that the measured isotope effect "which has recently come to my notice" "follows quantitatively" from his theory and "this agreement provides a direct check for the fundamental assumptions of the theory."

In other words, with these affirmations, repeated and amplified in numerous later articles by Fröhlich, Fröhlich was claiming that his theory had been developed independently and without knowledge of the experiments that measured the isotope effect, contrary to Bardeen's theory [4] that had been developed in response to these experiments.

In the next chapter, we will analyze in more detail these claims by Fröhlich. Continuing with the story, these events generated a fierce competition between Fröhlich and Bardeen. In the following years, both scientists continued their investigations based on the premise that the electron–phonon interaction is the origin of superconductivity. For example, in the following article that Bardeen sent for publication, in July 1950 [7], he mentions Fröhlich's name 18 times. In the following article published in 1951 [8], Bardeen compares in detail the theory of Fröhlich and his own, and concludes that neither explains in a convincing way how the electron–phonon interaction gives rise to superconductivity, but maintains the hypothesis that this is true. The article begins by saying "*The isotope effect, discovered independently by E. Maxwell, Reynolds et al., indicates that superconductivity arises from interactions between electrons and vibrations of the crystal lattice*". He also recognizes explicitly that "*Prior to his knowledge of the isotope effect, Fröhlich developed*" his theory.

Finally, in 1957 Bardeen, Cooper and Schrieffer published BCS theory [9], that explains superconductivity based on pairing of electrons due to the electron–phonon interaction, that is the theory generally accepted today. The article begins with the sentence "*A theory of superconductivity is presented, based on the fact that the interaction between electrons resulting from virtual exchange of phonons is attractive when the energy difference between the electrons states involved is less than the phonon energy*". The theory also predicts the isotope effect, that T_c varies proportionally to $M^{-1/2}$. Ever since then to the present, to doubt that the electron–phonon interaction gives rise to superconductivity is equivalent to saying that the Earth is not round but flat.

But what is the evidence that the critical temperature of a superconductor varies as $M^{-\alpha}$, with $\alpha = 1/2$, as Fröhlich 'predicted' in 1950, as Maxwell [1] and Reynolds *et al.* [2] found in Hg in 1950, and as BCS calculated in 1957?

Figure 12.1 shows the 'isotope coefficient' α measured for a variety of elements and compounds, from a paper by Phil Allen [10], an expert in the electron–phonon interaction, H-index 56, another 'little villain'. His most cited article, from 1975, where he calculates the transition temperature of various metals using the electron–phonon interaction, has 1673 citations. The article that contains Fig. 12.1 is from 1988, shortly after the cuprate superconductors were discovered, that clearly are not explained by the electron–phonon interaction. In that article, Allen reveals "*This electron–phonon mechanism is now universally accepted as a dominant cause of superconductivity in most known superconductors. Ironically, the isotope shift which provided the original proof of this mechanism is now one of the least well explained properties*".

Precisely. Figure 12.1 shows for example that for the elements *Ru* and *Zr*, α is zero. For *Os*, it is smaller than 0.2. How would

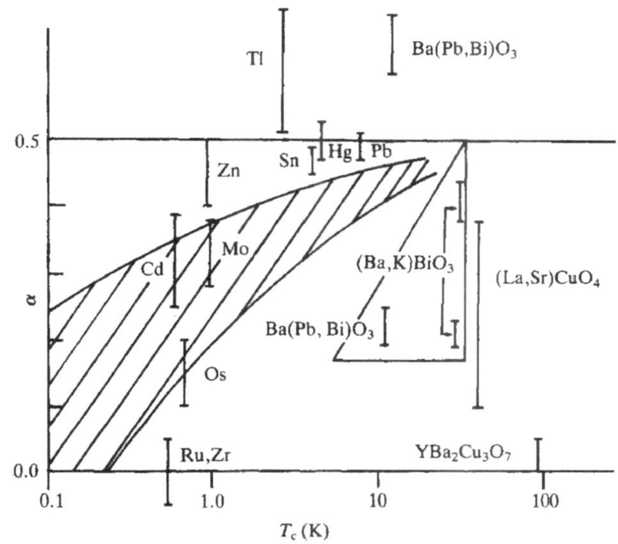

Fig. 12.1 Isotope coefficient α in the relation $T_c \propto M^{-\alpha}$, with M the ionic mass, measured in various experiments, from Ref. [10]. For the compounds in the figure, M is the mass of the oxygen atom.

the understanding of superconductivity have developed if in 1950 it had occurred to Maxwell [1] and Reynolds *et al.* [2] to measure the critical temperature of isotopes of Ru instead of isotopes of Hg?

Superconductivity misunderstood begins with Hg.

The zero isotope effect of Ru and Zr, and a *negative* isotope coefficient measured in Uranium, $\alpha = -0.53$ [11], not shown in Fig. 12.1, were some of the reasons for why Bernd Matthias doubted that BCS theory explains superconductivity in those elements.

Superconductivity properly understood begins with Heresy.

Another example that Allen mentions in his 1988 article as something that BCS can't explain is the compound HPd (palladium hydride), that shows an *increase* in the critical temperature, from 9 K to 11 K, when hydrogen is substituted by deuterium, therefore *negative* isotope coefficient. Note that the ionic mass here changes by a factor 2, instead of a factor 0.01 as in the experiments with mercury isotopes. Being the relative change in the mass 200 times larger than for the case of Hg, one would expect a clear agreement

Li	Be 0.026 ...	Superconductivity parameters for elements Transition temperature in Kelvin Critical magnetic field in gauss (10^{-4} tesla)										B	C	N	O	F	Ne ... ...
Na	Mg	...										Al 1.140 105	Si* 7 ...	P* 5 ...	S*	Cl	Ar ... ...
K	Ca	Sc	Ti 0.39 100	V 5.38 1420	Cr*	Mn	Fe	Co	Ni	Cu	Zn 0.875 53	Ga 1.091 51	Ge* 5 ...	As* 0.5 ...	Se* 7 ...	Br	Kr ... ...
Rb	Sr	Y*	Zr 0.546 47	Nb 9.50 1980	Mo 0.90 95	Tc 7.77 1410	Ru 0.51 70	Rh 0.0003 0.049	Pd	Ag	Cd 0.56 30	In 3.4035 293	Sn(w) 3.722 309	Sb* 3.5 ...	Te* 4 ...	I	Xe ... ...
Cs* 1.5 ...	Ba* 5 ...	La(fcc) 6.00 1100	Hf 0.12 ...	Ta 4.483 830	W 0.012 1.07	Re 1.4 198	Os 0.655 65	Ir 0.14 19	Pt	Au	Hg 4.153 412	Tl 2.39 171	Pb 7.193 803	Bi* 8 ...	Po	At	Rn ... ...

Fig. 12.2 Superconducting critical temperatures for the elements in K (number immediately below the chemical symbol). The * indicates the elements where superconductivity is under pressure or in a thin film, in a crystallographic structure that is not the most stable for the element.

with the theory that predicts that the critical temperature should decrease. But it is not so.

Superconductivity not understood begins with HPd.

I believe the clearest evidence that the electron–phonon interaction has nothing to do with superconductivity can be found in the periodic table. Consider Fig. 12.2. The mass of the ions increases when we go down and right in the periodic table. There is certainly no evidence in Fig. 12.2 that critical temperatures decrease in going from left to right and from up to down, as the formula $T_c \propto M^{-1/2}$ predicts.

But ok, one could argue that electronic properties change a lot between different columns of the periodic table, and this has effects that dominate over the dependence on ionic mass. Let's then look within one column, the valence electrons have the same quantum numbers (except for the principal one) for different elements in the column, for that reason elements in the same column have very similar chemical properties (that's the "periodic" in the periodic table!). The chemical properties derive from the valence electrons that are also the ones that have to do with superconductivity. Then, comparing the critical temperature of different elements in a given column is in a way equivalent to the experiment that measures isotope effect in an element. One would expect that the relation $T_c \propto M^{-1/2}$ would be satisfied at least approximately along a column. Is that so?

For example, Ti, Zr and Hf are in the same fourth column, with ionic masses 48, 91 and 179. According to the relation $T_c \propto M^{-1/2}$, given $T_c = 0.39\,\mathrm{K}$ for Ti, the T_c's for Ti, Zr and Hf should be $0.39, 0.28, 0.20$, not $0.39, 0.546, 0.12$ as the table shows. For V, Nb, and Ta in the fifth column, the T_c's should be $5.38, 4.00, 2.86$ instead of $5.38, 9.50, 4.48$ as the table shows. For Zn, Cd, and Hg they should be $0.875, 0.69, 0.34$ instead of $0.875, 0.56, 4.153$. Or from down to up, given $T_c = 0.14$ for Ir, T_c for Rh should be 0.19 instead of 0.0003; the T_c of Sn should be 9.48 instead of 3.72, given $T_c = 7.19$ for Pb.

In a word, there is *absolutely no evidence* in the periodic table that the relation $T_c \propto M^{-1/2}$ has anything to do with superconductivity. Fortunately, since otherwise, given the T_c of $0.875\,\mathrm{K}$ for Zn, the T_c of Hg in the same column would have been $0.34\,\mathrm{K}$ instead of $4.153\,\mathrm{K}$ and Kammerlingh Onnes would never have discovered superconductivity!

As a matter of fact, the question whether there was any correlation between ionic mass and critical temperature in the periodic table had been considered shortly before 1950, in 1947 [13]. The authors didn't find any simple correlation, certainly nothing resembling $T_c \propto M^{-1/2}$. The authors tried with complicated arguments to infer some prediction from these data, and predicted for example that Ce, Pr, and Nd should be superconducting (they had not yet been measured at that time). None of them is in their stable structures.

The lightest superconducting element according to Fig. 12.2 is Be, atomic weight 9, with $T_c = 0.026\,\mathrm{K}$. In fact, Fig. 12.2 is incomplete, in 2007 it was discovered that Li is superconducting at ambient pressure with critical temperature $T_c = 0.0004\,\mathrm{K}$. Li, with atomic weight 6.94, is the lightest superconducting element.

So the lightest superconducting elements, Li and Be, have extremely low T_c's, completely opposite to what the relation $T_c \propto M^{-1/2}$ suggests. According to that relation, if the critical temperature of Pb is $7.193\,\mathrm{K}$, the critical temperature of Li should be $39\,\mathrm{K}$ instead of $0.0004\,\mathrm{K}$, 100,000 times higher!

Does this make physicists doubt the veracity of this story? Does it make them think that perhaps there is a problem with the idea that the electron–phonon interaction causes superconductivity? Quite the contrary!

Since 1968 and especially in the last 10 or 15 years, there have been redoubled efforts to try to find superconductivity in the lightest element of all, hydrogen.

In 1968 Neil Ashcroft, one of the villains of Chapter 2, predicted that if one could make hydrogen metallic it would be superconducting at ambient temperature [14], due to the relation $T_c \propto M^{-1/2}$ and the light mass of H. Of course, hydrogen at ambient pressure is a gas, certainly doesn't conduct electricity. Oh well, we just have to apply sufficient pressure to make it into a metal, and under those conditions it will be a high temperature superconductor, said Ashcroft.

To make hydrogen metallic has turned out to not be easy, because one needs extreme pressures that have not yet been achieved. It doesn't matter, said Ashcroft in 2004 [15], alloys of hydrogen with other metals, as long as they are hydrogen-rich, should be sufficient to make superconductivity at high temperatures, and the pressures needed to make those systems metallic should be lower.

Ashcroft's 1968 article has 561 citations, half of which are in the last 7 years, 54 are in year 2018 alone. Ashcroft's 2004 article has 384 citations. It is a very current theme.

I will not enter into more details of this story. Suffice to say that in the last few years there have been enormous efforts devoted to this idea, enormous sums of money have been spent, and there have been several articles published in the last 5 years saying that at high pressure superconductivity at high temperatures has been experimentally detected in hydrogen-rich compounds [16]. However, none of these experiments has been independently reproduced in other laboratories to establish their validity. I believe they are not valid, these experiments are tricky and it is easy to make mistakes, and the theoretical expectations are conditioning these experiments in a way that they are probably not evaluated in a critical way.

Time will tell who is right.

Since we are talking about pressure, there is no reason to restrict ourselves to hydrogen. What happens if we apply pressure to other elements in the periodic table of Fig. 12.2? Following Ashcroft's reasoning, we would expect that T_c would be like that of metallic hydrogen predicted by Ashcroft, reduced by the factor $1/M^{1/2}$. Is it so?

Figure 12.3 shows critical temperatures of elements under pressure [17]. It can be seen that pressure has a large effect in increasing T_c in many cases. For example, Li under pressure superconducts at 14 K, a factor of 35,000 higher than at ambient pressure. Ca, that is not superconducting at ambient pressure, superconducts at 25 K under pressure, the highest critical temperature for an element under pressure. Other metals that are not superconductors at ambient pressure like Sc and Y superconduct under pressure at relatively high temperatures, 19.6 K and 19.5 K. Nb, that at ambient pressure superconducts at 9.5 K, the highest critical temperature for an element at ambient pressure, under pressure increases its critical temperature merely to 9.9 K. Other elements, like Zn and Pb, under pressure lower rather than increase their critical temperature.

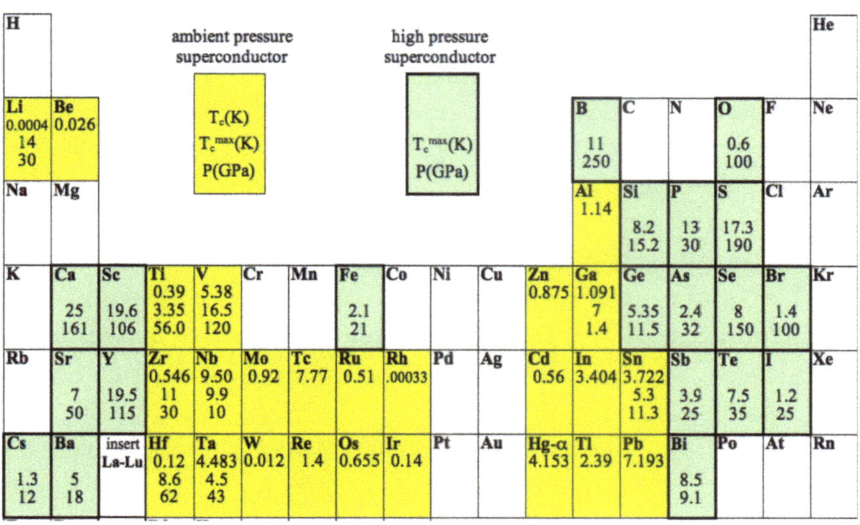

Fig. 12.3 Superconducting critical temperatures of elements under pressure, from Ref. [17]. In yellow, elements that are also superconductors at ambient pressure, in green those that only become superconducting under pressure. For each element, critical temperature is given in K and pressure in GPa.

Does the reader see any correlation in Fig. 12.3 between critical temperatures under pressure and ionic mass? I don't. If the reader doesn't either, does s/he think it is justified to devote enormous effort to search for high temperature superconductivity in hydrogen compounds based on the reasoning that hydrogen is the lightest element?

What do the superconductivity experts say about the behavior of critical temperatures under pressure? Not much. After T_c's are measured experimentally they try to explain them, using non-persuasive arguments about how the electron–phonon interaction and the electronic structure may change under pressure. They never predict them. In a rare moment of candor Yin, Savrasov and Pickett [18] admit that their complicated calculations that 'explain' the magnitude of T_c in some elements under pressure *after* measurements found what it is, "*has not yet provided — even for elemental superconductors — the physical picture and simple trends that would enable us to claim that we have a clear understanding of strong-coupling superconductivity.*"

Warren Pickett, another 'villain' apologist of the electron–phonon interaction, H-index 71. His second most cited article is from the year 2001, 'explaining' the 39 K superconductivity of the compound MgB_2 using the electron–phonon interaction, it has 674 citations. We will talk about that and other compounds in Chapter 15.

In summary, since 1950 it is universally believed that the electron–phonon interaction causes superconductivity. The evidence for this most frequently cited is the isotope effect. I will discuss another claimed evidence, within BCS theory, in Chapter 14. But based on what we saw in this chapter I maintain that the evidence in favor of the importance of the electron–phonon interaction based on the isotope effect is by no means convincing. Isotope effect occurs both in elements and compounds with magnitudes and signs different from what the theory predicts, frequently in relatively simple materials considered to be 'conventional' superconductors. After measured, complicated explanations are formulated, based on BCS plus effects of Coulomb repulsion and sometimes anharmonicity of vibrations, to explain the measurements. I argue that there can be many reasons unrelated to BCS's electron–phonon mechanism, for why the critical temperature may change with isotopic mass. I will mention one, within the theory of hole superconductivity, in Chapter 16. The transition temperatures in the periodic table of the elements, at ambient pressure and a high pressure, as well as evidence from hydrogen compounds and other compounds, does not show *any* tendency to higher critical temperature with lower ionic mass, as predicted by the theory of superconductivity based on the electron–phonon interaction.

From Wikipedia, "*Obscurantism is the practice of deliberately presenting information in an imprecise and recondite manner, often designed to forestall further inquiry and understanding. There are two historical and intellectual denotations of Obscurantism: (1) the deliberate restriction of knowledge — opposition to disseminating knowledge; and, (2) deliberate obscurity — an abstruse style characterized by deliberate vagueness*". All of these apply to this situation. It is nearly impossible to publish in scientific journals papers questioning the validity of the electron–phonon interaction to explain superconductivity, completely impossible in the highest prestige journals, due to opposition from referees and editors. The obscurantist period that began in 1950 persists today. My hope is that this book will speed up the advent of the Age of Enlightenment.

How did the science of superconductivity deviate into obscurantism? How were the profound physical ideas discussed in Chapter 11 forgotten? How was Aristarchus cast aside and replaced by Ptolemy? I believe what Herbert Fröhlich said, but much more what he *didn't say*, had a lot to do with it. That is the subject of Chapter 13.

References

[1] E. Maxwell, Isotope effect in the superconductivity of mercury, https://journals.aps.org/pr/abstract/10.1103/PhysRev.78.477.

[2] C. A. Reynolds, B. Serin, W. H. Wright, and L. B. Nesbitt, Superconductivity of isotopes of mercury, *Phys. Rev.* **78**, 487 (1950).

[3] J. Bardeen, Theory of superconductivity, https://journals.aps.org/pr/abstract/10.1103/PhysRev.108.1175.

[4] J. Bardeen, Zero-point vibrations and superconductivity, https://journals.aps.org/pr/abstract/10.1103/PhysRev.79.167.3.

[5] H. Fröhlich, Theory of the superconducting state. I. The ground state at the absolute zero of temperature, *Phys. Rev.* **79**, 845 (1950).
[6] H. Fröhlich, Isotope effect in superconductivity, *Proc. Phys. Soc. London* **A63**, 778 (1950).
[7] J. Bardeen, Wave functions for superconducting electrons, *Phys. Rev.* **80**, 567 (1950).
[8] J. Bardeen, Electron-vibration interactions and superconductivity, *Rev. Mod. Phys.* **23**, 261 (1951).
[9] J. Bardeen, L. N. Cooper, and J. R. Schrieffer, Theory of superconductivity, *Phys. Rev.* **108**, 1175 (1957).
[10] P. B. Allen, Isotope shift controversies, *Nature* **335**, 396 (1988).
[11] H. H. Hill *et al.*, Isotope effect in superconducting γ-Uranium alloys *Phys. Rev.* **163**, 356 (1967).
[12] B. Stritzker and W. Z. Buckel, Superconductivity in the palladium-hydrogen and the palladium-deuterium systems, *Z. Phys.* **257**, 1 (1972).
[13] J. De Launay and R. L. Doldeck, Superconductivity and the Debye characteristic temperature, *Phys. Rev.* **72**, 141 (1947).
[14] N. W. Ashcroft, Metallic hydrogen: A high-temperature superconductor?, *Phys. Rev. Lett.* **21**, 1748 (1968).
[15] N. W. Ashcroft, Hydrogen dominant metallic alloys: High temperature superconductors?, *Phys. Rev. Lett.* **92**, 187002 (2004).
[16] Emily Conover, A new hydrogen-rich compound may be a record-breaking superconductor, *Science News* **194**(7), 6 (2018).
[17] M. Debessai, J. J. Hamlin, and J. S. Schilling, Comparison of the pressure dependences of T_c in the trivalent d-electron superconductors Sc, Y, La, and Lu up to megabar pressures, *Phys. Rev. B* **78**, 064519 (2008).
[18] Z. P. Yin, S. Y. Savrasov, and W. E. Pickett, Linear response study of strong electron-phonon coupling in yttrium under pressure, *Phys. Rev. B* **74**, 094519 (2006).

Chapter 13

Herbert Fröhlich's deception and its consequences

On 16 May 1950, the journal *Physical Review* received the article by Herbert Fröhlich titled *"Theory of the Superconducting State. I. The Ground State at the Absolute Zero of Temperature"*, reference [5] in Chapter 12. In that article, Fröhlich proposed that the interaction between electrons and vibrating atoms in a solid can lead to a *redistribution* of electronic states in the space of momenta. The distribution in the normal state is simply a sphere in the space of momenta, all the states with momentum less than a maximum one (the radius of the sphere) are occupied by electrons, and those with larger momenta are all unoccupied. The new distribution proposed by Fröhlich is shown in Fig. 13.1, the dashed parts indicate occupied states. Here, states are occupied up to a maximum momentum $K_0 - A$, then there is a gap between $K_0 - A$ and K_0 where the states are empty, and then there is a ring of occupied states up to $K_0 + A$. Instead, in the normal state, the states up to K_0 would be occupied and those with momenta larger than K_0 would not.

Fröhlich presented calculations that purported to show that if the interactions between electrons and the vibrations of the solid are sufficiently large that distribution would have lower energy than the normal state, and that the difference in energy between that state and the normal state would be of the form

$$\Delta E = \frac{C}{M} \tag{13.1}$$

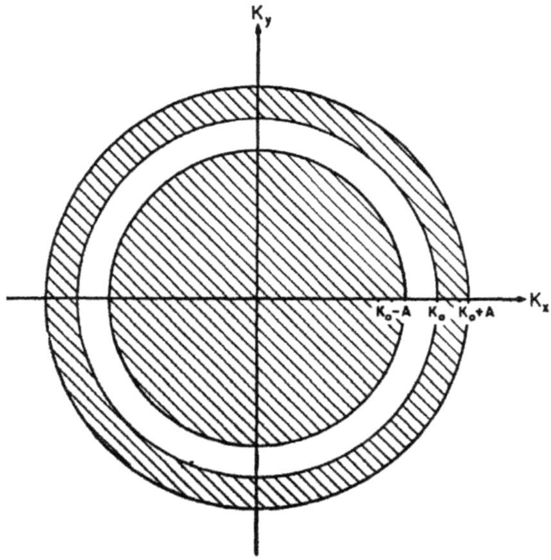

Fig. 13.1 Distribution of electronic states in the space of momenta in the superconducting state proposed by Fröhlich in 1950.

where C is a constant and M is the ionic mass. Besides, he presented complicated arguments that he said indicated that modifications of this state give rise to electric current that is not affected by normal resistance, and that this corresponds to the superconducting state that electrons adopt at low temperatures.

As motivation for this work, Fröhlich mentioned a paper he had written a year earlier about how an electron interacts with ionic vibrations in an ionic crystal [1], and said that this new work is the natural extension of the same formalism to the case of a metal. The article doesn't mention the experiments on isotope effect in superconductors by Maxwell and Reynolds *et al.* discussed in Chapter 12.

As we related in Chapter 12, three days after sending this article for publication Fröhlich sends a half-page letter to the english journal *Proc. Phys. Soc.* saying that he just found out about the measurements of isotope effect in mercury, and that his formula (Eq. (13.1) above) predicts exactly what is measured, $T_c \propto M^{-1/2}$, due to the fact that $\Delta E \propto T_c^2$.

During his entire life, Fröhlich maintained, in numerous scientific articles that he wrote, that he developed this calculation and wrote this article in May 1950 without having any knowledge of the experiments [2]. He relates it this way in an article that he wrote in 1983 [3]:

> "*The isotope effect in superconductors was in fact discovered during the same period. At that time I was at a 2–3 month visit to Purdue University, and submitted my paper on leaving on the 16 May 1950. I then spent a couple of days at Princeton and there, at my breakfast table, found the Physical Review with the two letters reporting the isotope effect* [12, 13]. *On checking I found my M-dependence confirmed and on the 19 May 1950 sent a letter* [14] *to claim confirmation of the basic idea, electron–phonon interaction.*"

That is, according to Fröhlich's version, these two events, (1) he sending his paper for publication that (implicitly) predicted the isotope effect, and (2) reading at his breakfast table two days later the articles that measured isotope effect, are *completely* independent events. What is the probability that these two independent events occurred separated by merely a couple of days, in the interval of 14,283 days since superconductivity was discovered (8 April 1911) until that moment?

The scientific world unanimously accepted Fröhlich's version of events. Why? Because it is unusual that scientists lie, so it is generally assumed that they tell the truth. In numerous scientific articles and books about superconductivity written from 1950 to the present, Fröhlich is given credit for having *predicted* the isotope effect and having *predicted* that the electron–phonon interaction is responsible for superconductivity without *any* knowledge of the experiments that measured the isotope effect in 1950. I don't agree with that interpretation, and published an article in 2011 [2] proposing a different version.

Continuing with the story of May 1950, after visiting Princeton Fröhlich visited Bardeen at Bell Laboratories where he worked, a few days after Bardeen had submitted his note explaining the isotope effect for publication (Ref. 4 in Chapter 12). During that encounter, Bardeen learned about Fröhlich's work, and was told by Fröhlich that he had done his work without any knowledge of the experiments. Bardeen believed it, and repeated this in many articles that he wrote, giving reference to Fröhlich's 1950 work and giving Fröhlich credit for having formulated his theory without knowledge of the experiments. Bardeen also told Fröhlich during that encounter about his own article motivated by the isotope effect experiments. From that point on, there was an intense competition between Bardeen and Fröhlich to understand in detail how the electron–phonon interaction causes superconductivity.

In this Fröhlich–Bardeen marathon that took place between 1950 and 1957, the advantage and the priority was Fröhlich's, because Fröhlich had 'predicted' the importance of the electron–phonon interaction, while Bardeen had 'postdicted' it. Bardeen's little article of May 1950 has a total of 71 citations in the scientific literature, 10 of which are self-citations as Bardeen was trying to promote his paper, while Fröhlich not even once cited it. In contrast, Fröhlich's article of May 1950 has 591 citations, Bardeen himself cited it 13 times.

But, Bardeen won. In 1957, BCS formulated their theory based on pairing of electrons induced by the electron–phonon interaction, and with the pairing concept they formulated the wave function Eq. (8.1) that, in fact, has very interesting properties that do explain some properties of superconductors. It was extended a few years later by the Russian physicist Eliashberg to describe in more detail the electron–phonon interaction. It is the theory universally accepted today for conventional superconductors.

On the other hand, Fröhlich's theory of May 1950, based on the momentum distribution of Fig. 13.1, became completely discredited shorty after published. For a variety of reasons, it makes no physical sense and it is mathematically inconsistent [4, 5], so much so that Fröhlich himself abandoned it shortly after publishing it. What he never abandoned though was his claim that he had discovered that the electron–phonon interaction causes superconductivity without knowing about the isotope effect experiments.

What sense does that make?

Let's assume BCS is correct in describing superconductivity as resulting from pairing of electrons through the electron–phonon interaction. Fröhlich didn't know anything about pairing (that physics was first introduced by Cooper in 1956), hence he did not have a valid theoretical basis that would lead him to deduce that the electron–phonon interaction causes superconductivity. So then there are three possibilities: either he got it right by complete accident, or he had divine clairvoyance, or *he knew about the experimental results* and formulated his wrong theory (Fig. 13.1) based on the electron–phonon interaction in an attempt to explain the experimental results, without saying so.

Clearly, the third possibility is the most plausible one. How could it have happened?

Fröhlich was at Purdue University that year from January to May. Shortly after arriving in the US, he could have learned through an informal channel about the initial results of these experiments, that were already obtained in January 1950. A likely source for that information would have been Karl Herzfeld, a collaborator of Emanuel Maxwell, possibly through Walter Heitler, a former student of Herzfeld and former collaborator of Fröhlich. Having Fröhlich just concluded a paper where he had studied the interaction of an electron with ionic vibrations in an ionic solid [1], it is logical that he would have wondered: if the critical temperature of superconductors varies with the ionic mass M the way these experiments are telling us, perhaps I could explain it using the formalism of electron–phonon interaction I was just studying for ionic solids. That way, he had several months to think about it and develop a possible explanation.

In fact, on March 20–21, 1950, there was a conference attended by more than 60 scientists [6] where Maxwell and Reynolds *et al.* announced the results of their measurements on isotope effect. A few days later, on 24 March, Maxwell and Reynolds *et al.* sent their papers for publication. It is not difficult to imagine that the information presented in that scientific meeting reached Fröhlich, almost two months before he completed and sent out his paper for publication.

If Fröhlich had said that his 1950 theory was formulated to explain the results of the isotope effect experiments, his work would not have had much impact, because his theory was no good. Fröhlich's 1950 article had a very big impact because supposedly his work was performed without knowledge of the experiments. That is the crux of the matter.

In physics (and other sciences), it is much easier to 'explain' than to 'predict'. Once something has been measured, it is easy to imagine a variety of possible explanations. There are few situations where exact calculations are feasible, in general calculations are approximate, the parameters that enter the calculation are not exactly known and cannot be measured, and it can be difficult to discern which explanation is correct among various alternatives. A clear example is superconductivity in the cuprates. Innumerable different explanations for the phenomenon have been proposed, and there is no agreement whether any of them is correct. Imagine that somebody would have *predicted* high temperature superconductivity in the cuprates before it was discovered. Clearly, that explanation would have *much* more credibility than explanations proposed *a posteriori*.

So let us imagine that Fröhlich would *not* have 'predicted' the isotope effect. The hypothesis that the electron–phonon interaction causes superconductivity would not have had the acceptance that it had. Even more so if during that time measurements of zero and negative isotope coefficients would have been made, that as we mention exist in other elements. Perhaps BCS would have formulated their pairing theory and their wave function Eq. (8.1) with a different interaction leading to pairing of electrons, there are many processes analogous to the electron–phonon interaction that can do it. For example, polarizability of some electrons can mediate an attractive interaction between other electrons that leads to pairing. Such mechanisms have been proposed to explain the superconductivity of the cuprates.

If history had developed that way, and if in addition some of the many materials that are today considered to be 'unconventional' had been found back then, the development of the understanding of superconductivity would have been very different. Probably the search for a *single* mechanism of superconductivity that explains in a unified way all superconducting materials would have continued, as in the early days of superconductivity.

But history did not develop that way, in large part, I argue, because of Fröhlich's deception, that led the scientific world to be fooled into thinking that the electron–phonon interaction as the root cause of superconductivity is something logical, inevitable, and irresistible, as demonstrated by the fact that it could be predicted without doing experiments. Fröhlich's dishonesty led the scientific world to wander for many years (until today) in the desert of multiple explanations of superconductivity, different for 'conventional' and 'unconventional' superconductors, unable to reach the promised land of superconductivity properly understood.

Anyway, thanks to Fröhlich, or more appropriately due to Fröhlich's fault, history developed the way it did. 1957 was the beginning of the 'golden age' of BCS theory, that lasted until approximately 1980, as we will see in Chapter 14.

References

[1] H. Fröhlich, H. Pelzer, and S. Zienau, Properties of slow electrons in polar materials, *Phil. Mag.* **41**, 221 (1950).

[2] J. E. Hirsch, Did Herbert Fröhlich predict or postdict the isotope effect in superconductors?, *Phys. Scr.* **84**, 045705 (2011).

[3] H. Fröhlich, History of the theory of superconductivity, in *Advances in Superconductivity*, ed. by B. Deaver and J. Ruvalds (New York: Plenum) p. 1 (1983).

[4] G. Wentzel, The interaction of lattice vibrations with electrons in a metal, *Phys. Rev.* **83**, 168 (1951).

[5] W. Kohn and Vachaspati, A difficulty in Fröhlich's theory of superconductivity, *Phys. Rev.* **83**, 462 (1951).

[6] W. T. Ziegler, ONR cryogenics conference, *Science* **111**, 525 (1950).

Chapter 14

1957–1980: The golden age of BCS theory

At the same time that BCS theory was published in 1957, the US newspaper New York Times published a story about it, titled "Theory of metals in cold evolved", that began by saying "*A new mathematical theory appears to provide for the first time an explanation for the half-century old phenomenon of superconductivity. It is being pondered by physicists around the world. Full details of the concept will be published in the current issue of The Physical Review, journal of the American Physical Society. The issue, dated Dec. 1, is now in the mails.*"

It is very unusual that a daily newspaper, especially of the caliber of the New York Times, writes about a scientific theory at the same time as it is published in a scientific journal. This was surely due to the fame that John Bardeen had, having received the 1956 physics Nobel prize (together with Brattain and Shockley) for the invention of the transistor. Certainly this must have contributed to the quick dissemination and acceptance of the theory.

The BCS theory attracted scientists' attention very rapidly. Already during the first year after its publication, i.e. 1958, 36 articles were published citing the BCS paper. In the first 5 years, till 1962, 498 published articles cite BCS. Figure 14.1 shows the annual number of published articles citing BCS from 1958 to 2018, a total of 7,866.

The paper by BCS (Ref. [1] in Chapter 10) has an entire section, V, 6 pages long, devoted to the Meissner effect. However, as we

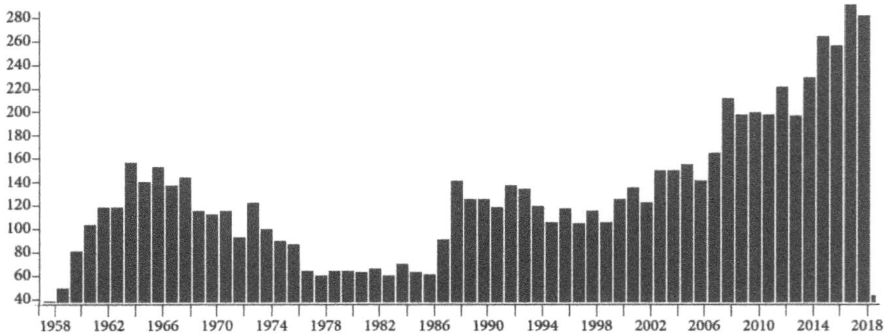

Fig. 14.1 Yearly number of articles citing the BCS 1957 article.

discussed in Chapter 8, it ignores the most important question: how does the normal metal expel the magnetic field in the process of becoming superconducting?

Nobody criticized that aspect of BCS at that time nor afterwards, what was initially criticized was that the theory and in particular the description of the Meissner effect did not satisfy what is known as 'gauge invariance', a mathematical property that valid physical theories must respect. This was pointed out and discussed in several scientific papers in the years that followed, but rapidly the conclusion was reached that by extending BCS's treatment it was possible to satisfy this restriction without altering BCS's conclusions.

In BCS's article and in subsequent years, BCS theory was used to calculate various properties of superconductors, and it was found that it reproduced many measured experimental results, for example thermodynamic, optical, acoustic, electric and magnetic properties. New experimental techniques were developed, e.g. tunneling, that allowed for the density of electronic states to be measured in superconductors, and verify that it coincided with what BCS predicted. Of course in all of these calculations it was necessary to put in parameters in the theory that were not calculated from first principles but adjusted to experiments.

Tunneling turned out to be a very important technique to study superconductivity. In these experiments, the surface of the superconductor is covered with a thin film of an insulating material, which creates a potential barrier, and a current is circulated from a normal electrode through this film into the superconductor. The quantum effect known as 'tunneling' allows electrons to pass through this barrier, and when they enter the superconductor the electrons 'sense' what is called the 'density of states' accessible to them; the conductance of the junction is proportional to the density of states. BCS theory had a very specific prediction, namely an energy gap and a peak in the density of states at the edge of this gap.

Tunneling experiments reproduce this prediction of BCS with remarkable accuracy. Figure 14.2 shows the measured conductance with a full line, and the calculated one with a dashed line for the case of Sn [1]. Its author, I. Giaever, received the physics Nobel prize in 1973 for these measurements.

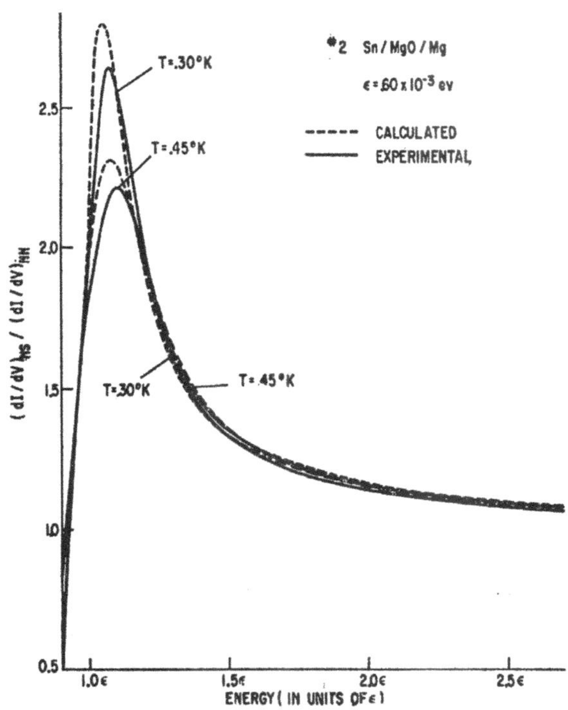

Fig. 14.2 Conductance of a tunnel junction with the superconductor Sn, as function of the voltage (or energy) across the junction. Note that the origin of coordinates is shifted to better show the details of the curves.

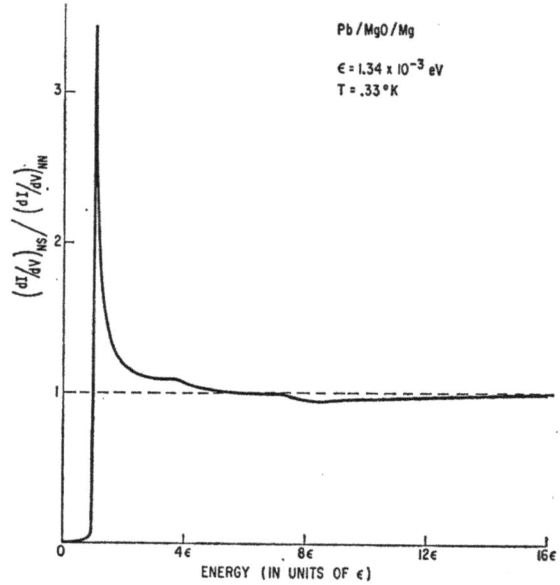

Fig. 14.3 Conductance of a tunnel junction with the superconductor Pb, as a function of voltage across the junction. Here the origin of coordinates is not shifted like in Fig. 14.2. Note the small oscillations in the curve near energies 4ϵ and 8ϵ.

But in addition it was found that in some metals there are small deviations from BCS. Figure 14.3 shows tunneling results for the superconductor Pb [2]. It can be seen that there are small oscillations at energies higher than the energy gap ϵ. BCS instead predicts a curve without oscillations as in Fig. 14.2.

The small oscillations in Fig. 14.3 caused great excitement in the scientific community, because they were interpreted as due to the vibrations of atoms that according to BCS give rise to superconductivity [3]. Elaborate calculations were performed using a mathematical formalism introduced by the Russian physicist Eliashberg [4] to take into account the dependence of the energy gap $\Delta(E)$ on the vibration frequency of the ions. In BCS theory, it is assumed for simplicity that Δ is a constant. With these calculations, it was shown that it was possible to reproduce the small oscillations in Fig. 14.3 [5–7]. These calculations convinced the scientific community that this theory is irrefutable. The small oscillations in Fig. 14.3

were considered the 'smoking gun', even more so than the isotope effect, that established that the electron–phonon interaction is the cause of superconductivity. Two of the most important participants in these efforts were Anderson [3] and Scalapino [5, 8], mentioned in Chapter 2.

I will explain in some more detail the abstruse reasoning of these scientists. They devised a calculation process called 'inversion' [6]. The tunneling conductance is measured, and through a complicated mathematical procedure one obtains a quantity called $\alpha^2(\omega)F(\omega)$, where ω is the vibration frequency of the ions. The function $F(\omega)$ is the spectrum of vibrations of the ions that is in principle measurable by other methods, e.g. neutron scattering. The function $\alpha(\omega)$ is the magnitude of the electron–phonon interaction that cannot be measured and also cannot be calculated with any precision. The calculation also has another parameter, the Coulomb interaction U_c (also called μ^*), that is not calculated but its value is guessed. Comparing the tunneling measurements that give $\alpha^2(\omega)F(\omega)$ with the neutron measurements and assuming a value of U_c the function $\alpha(\omega)$ is extracted. This is how the creators of this method explain it [6]: "*The fact that one can fit the tunneling data with reasonable values of $\alpha^2(\omega)F(\omega)$ and U_c provides a confirmation of the theory of superconductivity*".

But it is not so. The problem is, what does 'reasonable' mean? The reality is that this formalism doesn't confirm anything, because there is no independent way to determine neither the function $\alpha(\omega)$ nor the parameter U_c.

It is also a fact that this procedure was applied in detail to a *single* element, Pb, where an approximate consistency was found between what was deduced from tunneling measurements regarding the vibration spectrum of the ions, and the measurements of ionic vibrations with neutrons [8], without a need to assume large variations in the function $\alpha(\omega)$. In other materials, $\alpha^2(\omega)F(\omega)$ was obtained from tunneling measurements but the consistency with neutron experiments was not investigated [7, 9], or it was and no consistency was found [10]. And in other superconductors, one doesn't find in tunneling oscillations of the magnitude necessary to explain the observed T_c, unless one assumes $U_c < 0$, that is a Coulomb interaction between electrons that is attractive instead of repulsive [11]. Or, one has to invoke other effects, for example that the tunnel barrier is of low quality [12]. In my opinion, even assuming that the origin of the oscillations seen in the curve of Fig. 14.3 is the lattice vibrations, this by no means *proves* that the electron–phonon interaction causes superconductivity, just like the isotope effect doesn't prove it. Obviously there is always a coupling between electrons and lattice vibrations, that can manifest itself in different ways. For example, in semiconductors there is an isotope effect in the energy gap [13], and certainly in semiconductors we know that the gap *does not* originate in the electron–phonon interaction.

In any event, BCS theory and its extension due to Eliashberg, consolidated during the 60s as the established and confirmed theory of superconductivity, that was believed to apply to all superconducting materials. In 1972, Bardeen, Cooper, and Schrieffer received the Nobel prize for their theory. The press release of the Royal Swedish Academy said "*A significant step forward was taken around 1950 when it was found theoretically and experimentally that the mechanism for superconductivity had to do with the coupling of electrons to the vibrations of the crystal lattice. Starting from this mechanism, Bardeen, Cooper and Schrieffer developed in 1957 a theory of superconductivity, which gave a complete theoretical explanation of the phenomenon.*"

Case closed.

Or maybe not.

One would expect that if there is a theory that explains superconductivity, it should be possible to explain the values of critical temperatures that are found in the different materials, and understand which properties of materials are conducive to high critical temperatures. Knowing this, one can search and find new materials with higher critical temperature than those known. But none of this happened.

The first attempt to calculate the critical temperature of elements using BCS was carried out by David Pines, collaborator of

Bardeen, in 1958 [14]. Pines calculated the electron–phonon interaction by an approximate method and obtained a criterion for superconductivity that depends on a parameter he calls Z^*, the effective valence of the ions, and on the electron density denoted by r_s. In a graph with coordinates Z^* and r_s, he draws a curve that according to his calculation should separate the superconducting elements (above the curve) from those that are not (below the curve) as Fig. 14.4 shows.

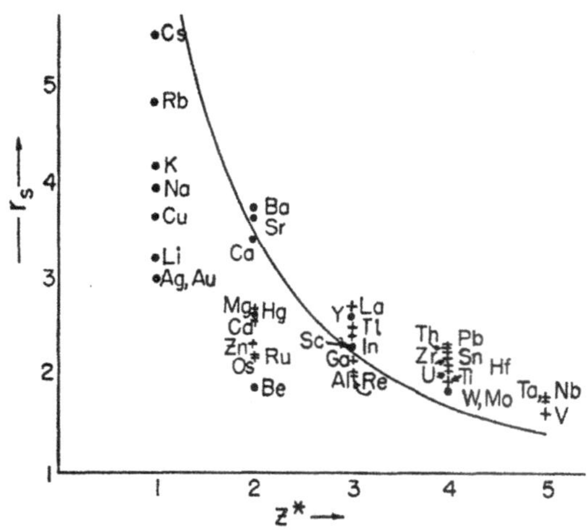

Fig. 14.4 The curve in the figure separates elements that Pines' theory [14] predicts are superconducting (above) from those that are not (below).

According to this graph, Ba, Sr, Y, and Sc should be superconductors, but are not. Hg, Zn, Ru, and Be, shouldn't be superconductors, but they are. Pines also tried to estimate values of T_c from his calculation, but concluded that the values for the electron–phonon interaction that he obtained were not large enough to yield T_c values that coincided with the experimental results.

Pines' work foreshadowed the difficulties that this theory has in calculating critical temperatures. Calculation methods became much more sophisticated in subsequent years, and it is not easy to understand the approximations that the authors make. The problem with these highly complex and sophisticated calculations is that since the answer is known in advance, there is an incentive to make approximations that yield the correct answer and not make approximations that don't. Let's mention some examples.

For example, Carbotte and Dynes in 1967 calculated [15] T_c for Al and found $T_c = 1.17\,\mathrm{K}$, very close to the measured value $T_c = 1.18\,\mathrm{K}$. They don't give many details, so that it is impossible for the reader to verify this result. Allen and Cohen in 1969 calculated [16] the critical temperature of 19 metals. For some, like Pb, Sn, Tl, Hg and Zn, the calculated T_c differed from the measured one by less than a factor of 2. For Ga, they found it differed by a factor of more than 20. For Li and Mg, they found they should be superconductors, and the authors urged that these materials be investigated, saying "*The discovery of superconductivity in these materials would be a rather convincing demonstration that the theory of the transition temperature had come of age.*" But as of today, no superconductivity has been found in Mg, and the $T_c = 0.0004\,\mathrm{K}$ in Li that was found is 2000 times smaller than what Allen and Cohen predicted, 1 K. Note that Li has only three electrons, it should be relatively easy to study it using Allen and Cohen's method. In another study, Papaconstantopoulos and coworkers in 1977 calculated [17] T_c for the 32 metallic elements with atomic number up to 49. Many of the obtained values were very different from the measured ones. For example, for In 0.04 K instead of 3.40 K, for Ru 0 instead of 0.49, for Sc 0.51 K instead of 0. For Al, for which Carbotte and Dynes had found T_c with an error of less than 1% 12 years earlier, Papaconstantopoulos *et al.* with more sophisticated techniques find $T_c = 0$ instead of the experimental value 1.17 K. Despite these discrepancies, the authors [17] concluded that their calculation was "*a promising step in the direction of predicting new superconductors in more complex materials.*"

The H-index of these authors? Carbotte H = 54, Dynes H = 56, Allen H = 56, Cohen H = 122, and Papaconstantopoulos H = 48. All high, indicating they are recognized and influential scientists.

These and many other works indicate that the calculation of critical temperatures even for the simplest materials, the elements at ambient pressure, using the BCS–Eliashberg theory, don't seem to be capable of predicting the T_c's that are measured. An expert in the field, D. Rainer, wrote in 1982, "*We must at least consider the possibility that although the present theory of T_c is able to explain every known experimental result, it is nevertheless unable to make reliable predictions.*" He called this situation "superflexibility, manifested in "*ad hoc assumptions, plausibility arguments, or semi-empirical rules*" that "*leaves too much room for uncontrolled enthusiasm.*"

Paraphrasing Rainer: knowing the result that has to be obtained, in complicated calculations that necessarily require to make approximations, to decide between alternative criteria, etc., it is possible to do it so as to obtain a result that is close to the experimental result. Or, it is possible to do it so as to obtain any other result.

Bernd Matthias, about whom we talked in Chapter 2, the only 'heretic' in this golden age of BCS, wrote in 1971 [19]: *Especially since 1957, with the advent of the Bardeen–Cooper–Schrieffer (BCS) theory many hundreds of papers and learned treatises have appeared, describing and predicting superconductivity at elevated temperatures, at room temperature, and even above. And yet, these papers have not led to a single success in raising the transition temperature. The deluge of idle speculations coming to us these days from all sides just won't do it — all it will manage to do is to widen the credibility gap instead of the energy gap. In the spirit of our times, there is an increasing tendency to substitute for nonexistent results many words of great expectations.*

But, the great majority of scientists, including the most prominent ones, certainly don't see it that way. Marvin Cohen, one of the 'villains' mentioned in Chapter 2, has devoted great efforts over many years, since 1964 and continuing to the present, in calculating T_c and other properties of superconductors, which earned him a good part of the 64,382 citations that his papers have, and his extremely high H = 122. Cohen maintains that he has predicted new superconductors before they were discovered, but it is not easy to find evidence that this is actually so. In an article he published in 2010, titled "Predicting and explaining T_c and other properties of BCS superconductors" [20], Cohen claims about conventional superconductors, which he calls 'class 1': "*A theoretical description of their superconducting properties based on electron pairing induced by electron–phonon coupling works extremely well for class 1, and it is possible to use the theory to predict new superconductors.*" But then he himself cautiously concedes: "*calculations after an experimental discovery are easier since it is very helpful to know the possible range of parameters involved and to know the answer. When a new candidate system is chosen, there are often many unknown variables to consider.*" In a court of law, such a statement would be excluded due to the fifth amendment.

In any event, the period between 1957 and the end of the 70s can rightly be considered the 'golden age' of BCS theory. There was complete confidence (with the exception of Matthias) that BCS theory with Eliashberg's extension (1) is completely correct, and (2) applies to *all* superconducting materials, and that eventually, once it was better understood how to calculate properties of normal metals with greater accuracy, it would be possible to calculate all the superconducting properties of known materials and materials to be discovered, including their T_c's, with this theory. The belief (1) continues today, the belief (2) no longer, as we will see in what follows.

The highest critical temperature for superconductivity at that time was $T_c = 23\,\mathrm{K}$ for the compound Nb_3Ge, discovered in 1973, member of the family of compounds known as A15. Interest in this family of compounds had started in 1954 with the discovery of superconductivity in Nb_3Sn with $T_c = 18\,\mathrm{K}$. That is, in over 20 years the maximum T_c had increased by a mere $5\,\mathrm{K}$. It was generally believed that it was impossible that the electron–phonon interaction could give rise to critical temperatures above approximately $25\,\mathrm{K}$, as was articulated in a famous article by Marvin Cohen and Phil Anderson (two of the villains of Chapter 2), in their paper titled "*Comments on the Maximum Superconducting Transition Temperature*" [21] published in 1972.

But then [22], in the mid-70's new superconductors were found, barium–lead–bismuth–oxides (BaPbBiO) that, even though they didn't have a very high transition temperature ($T_c \sim 13\,\mathrm{K}$) it seemed impossible to explain it with the electron–phonon interaction because the electronic density of states was too small. Then, at the end of the 70's two new classes of superconducting materials were found, organic superconductors and so-called 'heavy fermion superconductors', that also did not seem to conform to BCS theory. In these materials, the problem was not the critical temperature, that was not high at all, rather that certain properties like the specific heat at low temperatures appeared to indicate that there was no energy gap as predicted by BCS. The term 'unconventional superconductor' started to be used and popularized, characterizing superconductors that did not obey BCS theory and for which another theory had to be found to describe them. But it wasn't until 1986 that the subject of unconventional superconductors 'exploded' with the discovery of a copper oxide material that superconducted at 35 K, thus beating the record that until then was held by Nb_3Ge, after which rapidly other 'cuprates' followed with much higher critical temperatures. We continue with that topic in Chapter 15.

References

[1] I. Giaever, *Phys. Rev. Letters* **5**, 147 (1960).
[2] I. Giaever, H. R. Hart Jr., and K. Megerle, *Phys. Rev.* **126**, 941 (1962).
[3] J. M. Rowell, P. W. Anderson, and D. E. Thomas, Image of the phonon spectrum in the tunneling characteristic between superconductors, *Phys. Rev. Lett.* **10**, 334 (1963).
[4] G. M. Eliashberg, *Soviet Phys. JETP* **11**, 696 (1960).
[5] J. R. Schrieffer, D. J. Scalapino, and J. W. Wilkins, *Phys. Rev. Lett.* **10**, 336 (1963).
[6] W. L. McMillan and J. M. Rowell, *Phys. Rev. Lett.* **14**, 108 (1965).
[7] W. L. McMillan, *Phys. Rev.* **167**, 331 (1968).
[8] D. J. Scalapino, in "Superconductivity", edited by R. D. Parks, Marcel Dekker, Inc., New York, 1969, Vol. I. p. 449.
[9] W. L. McMillan and J. M. Rowell, in "Superconductivity", edited by R. D. Parks, Marcel Dekker, Inc., New York, 1969, Vol. I. p. 561.
[10] W. Weber, *Physica B+C* **126**, 217 (1984).
[11] J. Bostock *et al.*, *Phys. Rev. Lett.* **36**, 603 (1976).
[12] G. B. Arnold, J. Zasadzinski, and E. I. Wolf, *Phys. Lett. A* **69**, 136 (1978).
[13] E. E. Haller, *Solid State Commun.* **133**, 693 (2005).
[14] D. Pines, *Phys. Rev.* **109**, 280 (1958).
[15] J. P. Carbotte and R. C. Dynes, *Phys. Lett. A* **25**, 685 (1967).
[16] P. B. Allen and M. L. Cohen, *Phys. Rev.* **187**, 525 (1969).
[17] D. A. Papaconstantopoulos *et al.*, *Phys. Rev. B* **15**, 4221 (1977).
[18] D. Rainer, *Physica B* **109** and **110**, 1671 (1982).
[19] B. T. Matthias, *Physics Today* **24**(8), 23 (1971).

[20] M. L. Cohen, *Modern Physics Letters B*, **24**, 2755 (2010).
[21] M. Cohen and P. W. Anderson, in "Superconductivity in d- and f-Band Metals", ed. by D.H. Douglass (AIP, New York, 1972), p. 17.
[22] For a survey of superconducting materials classes see *Physica C* Special Issue: "Superconducting Materials: Conventional, Unconventional and Undetermined", edited by J. E. Hirsch, M. B. Maple and F. Marsiglio, *Physica C* 514, p. 1–444 (2015).

Chapter 15

Holes in cuprates and other materials

In 2006, commemorating BCS theory, the American Physical Society wrote [1]: "*In 1957, the BCS theory explained low-temperature superconductivity almost 50 years after its discovery.*" It characterizes BCS as "*one of the most successful theories in solid state physics*" and adds: "*The BCS theory also had an important influence on theories of particle physics and provided the starting point for many attempts to explain the new high-temperature superconductors.*" I want to emphasize the last sentence: "*provided the starting point for many attempts to explain the new high-temperature superconductors.*" That's precisely the problem. During the last 30+ years, since the discovery of the cuprates in 1986, physicists have been trying to explain high temperature superconductivity 'starting' from BCS theory, or at least assuming implicitly that BCS theory is correct for the low temperature superconductors [2]. That is what I say led to the swamp in which we are today. To understand 'unconventional superconductors', it is necessary to question the understanding of 'conventional superconductors'.

The era of high temperature superconductivity started in September 1986, when Bednorz and Müller published an article cautiously titled "*Possible high T_c superconductivity in the Ba–La–Cu–O system*", where they informed that in that material superconductivity appeared to start around 35 K, a much higher temperature than the maximum T_c of 23 K known. A few months later, Paul Chu (mentioned in Chapter 2) showed that in the system YBaCuO superconductivity occurs around 90 K. This was a revolution, particularly because now T_c was above 77 K, the temperature at which nitrogen becomes a liquid at ambient pressure, which makes experiments much easier and less costly. Bednorz and Müller's article has 9,832 citations, Chu's article 5,189.

What all cuprate superconductors have in common are planes that contain copper and oxygen in the configuration shown in Fig. 15.1. Between the planes, there is a variety of other atoms depending on the material.

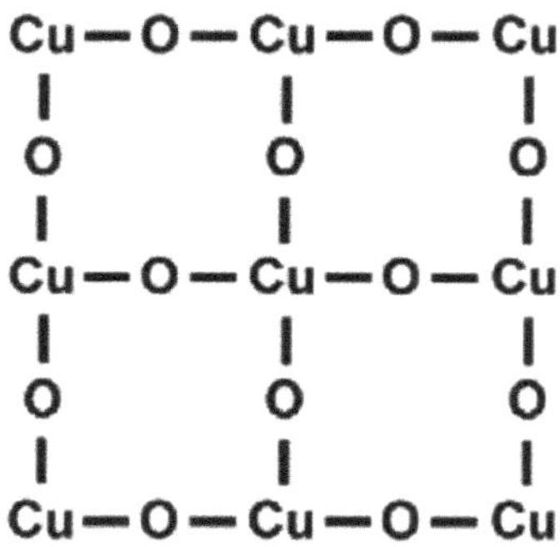

Fig. 15.1 Copper–oxygen planes that give rise to high temperature superconductivity, for reasons that are as yet not understood by the scientific community.

The history of these initial months and years of 'high T_c', high temperature superconductivity, is very interesting, but we will not enter into much detail here. What is important is to point out that since the very first moment, and certainly until today, nobody had doubt that these materials could not be explained by BCS theory: there is no way the electron–phonon interaction can give rise to superconductivity at such high temperatures. The scientific community turned toward searching for "the mechanism" that would explain superconductivity in the class of materials discovered by Bednorz and Müller, the so-called 'cuprates'.

The attention of most physicists, from the beginning till today, centered in the *magnetic properties* of the material discovered by Bednorz and Müller. This was to a great extent motivated by an article published by P. W. Anderson in early 1987 with title "The Resonating Valence Bond State in La_2CuO_4 and Superconductivity", where he proposed that the normal state of these materials is not a normal metal but a 'spin liquid', that already has paired electrons due to magnetic interactions, but does not yet have the phase coherence that characterizes a superconductor. He proposed the Hubbard model, mentioned in Chapter 4 and about which we will talk more later, to characterize this state. This article by Anderson had enormous impact, it has today 6002 citations, in turn those 6002 papers have been cited 249,748 times in other articles. Scientists became enamored with the 'spin liquid' concept, even though even to this day there is no clear evidence that such a state exists in nature, let alone in La_2CuO_4 nor in any other superconducting material. There have been innumerable proposals of unconventional mechanisms of superconductivity based on this concept, and this has conditioned the evolution of this field till the present. It could be an interesting theme for a future book, titled for example 'Tulip mania, spin liquids and superconductivity', but it is not the subject that occupies us in this book.

It didn't take very long since Bednorz and Müller's discovery till scientists started to realize that *holes* in these materials play a special role. The first paper calling attention to this fact was by the Japanese physicist Shin-ichi Uchida (H-index 81) and coworkers [3], submitted for publication 14 March 1987, published in April 1987. Analyzing the electrical transport properties of the compound $(La_{1-x}Ba_x)_2CuO_4$ as function of the barium (Ba) concentration (x), they deduced that "*holes have dominant contribution to the conduction*". Soon thereafter, in May 1987, theorist V. Emery submitted for publication a paper titled "*Theory of high-T_c superconductivity in oxides*" where he formulated and analyzed a model [4] where the charge carriers were holes in the $2p$ states of oxygen. Soon thereafter, experimentalists M. W. Shafer and coworkers submitted for publication a paper [5] titled "*Correlation of T_c with hole concentration in $La_{2-x}Sr_xCuO_{4-\delta}$ superconductors.*" At the low temperature physics conference in Kyoto in August 1987, a large number of presentations were focusing on holes. At that meeting, Bednorz and Müller presented a paper titled "*A Road towards High T_c Superconductivity*" where they pronounced "*Basically, all these materials, are hole superconductors*" [6].

Thus began what I call the 'era of hole superconductivity'. Before 1987, there had hardly ever been mention of 'holes' in connection with superconductivity. Figure 15.2 shows a graph of the yearly number of published papers on superconductivity with the word 'hole' in the title. Practically none before 1987, during the previous 75 years since superconductivity was discovered in 1911. By contrast, Fig. 15.3 shows the same for articles on semiconductors with the word 'hole' in the title. For semiconductors, from the moment they started to be used in electronic devices around 1950, the concept of 'holes' played an important role. This is also illustrated by the title of the famous

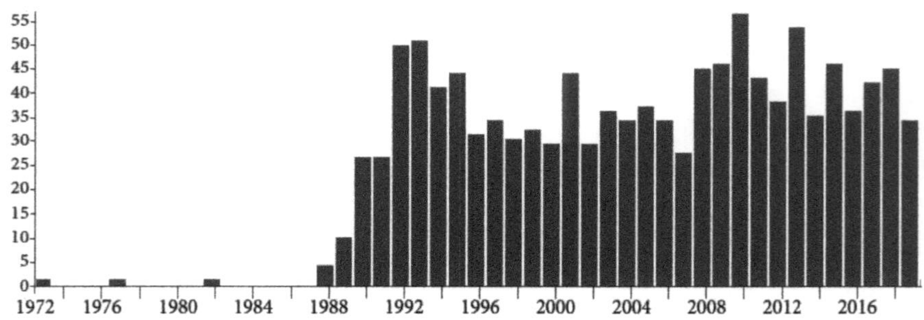

Fig. 15.2 Number of yearly articles on superconductors with the word 'hole' in the title.

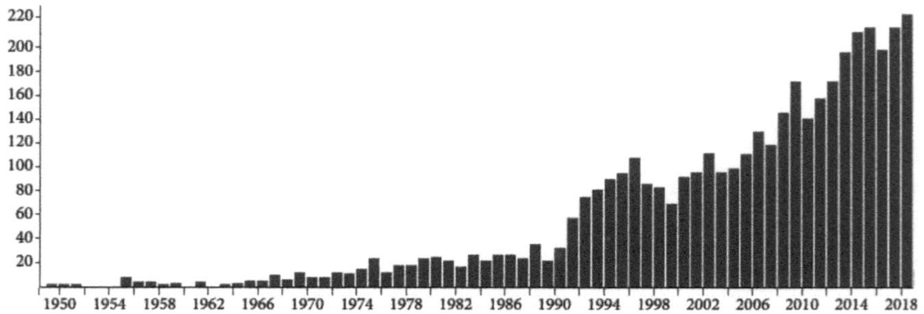

Fig. 15.3 Number of yearly articles on semiconductors with the word 'hole' in the title.

1950 book by Shockley (1956 Nobel laureate for the invention of the transistor), *Electrons and Holes in Semiconductors.*

It would be logical to think, given that the great increase in the critical temperature of superconductors coincided with the realization that holes play an important role in the cuprates, that physicists would have concluded that holes are important for superconductivity. Surprisingly, that didn't happen. Ironically, this was in large part because of an article that soon thereafter Uchida himself published, the same researcher who had first called attention to the importance of holes in the cuprates.

On 26 January 1989, the prestigious journal *Nature* published an article titled "*A superconducting copper oxide compound with electrons as the charge carriers*" [7] by the Japanese physicists Tokura, Takagi, and Uchida. In compounds with chemical formula $A_{2-x}Ce_xCu0_{4-y}$, with A the element Nd, Pr, or Sm, the authors found that the charge carriers in the planes were electrons instead of holes. These compounds have the same planar structure shown in Fig. 15.1 but different atomic configuration off the planes. This seemed to indicate that there is nothing special about holes to give rise to superconductivity. Enamored of the concept of electron–hole symmetry, that is how this finding was interpreted by the majority of physicists, even to this day. The physical chemist Art Sleight, expert in these materials, commented upon this discovery [8]: "*This symmetry between adding and subtracting electrons will have to be reflected in any theory that explains high-temperature superconductivity, and existing theories based on the supposition that there is something unique about hole carriers are 'out the window'.*"

One of these "existing theories" was the theory that I and collaborators had started to develop only a few months earlier. My first paper with this point of view, titled *Hole superconductivity* [9], had just been published 3 days earlier, 23 January 1989. That is, it existed exactly 3 days until being thrown "out the window" by the Tokura, Takagi, and Uchida discovery. A record of ephemerality.

Or not?

In that article, I proposed that 'holes' explained the high temperature superconductivity in cuprates, and in addition played an essential role in the superconductivity of *all* materials. No compromise was possible. Either my paper was wrong, or Tokura, Takagi, and Uchida's paper was wrong. Their paper has 1,476 citations today, mine only 94.

Even so, Tokura, Takagi, and Uchida's paper was wrong.

What these scientists had done was to observe that substituting (for example) Nd atoms by Ce atoms in the compound, the conductivity increased. Taking into account the valence of Nd and Ce, this substitution implied that electrons were being added to the compounds. For that reason, those materials are known as 'electron-doped cuprates'. They had also measured the Hall effect (remember that the Hall coefficient is negative if the charge carriers are electrons, positive if they are holes, see Chapter 5) and obtained a negative result. Logically, they inferred that the charge carriers were electrons. But they got ahead of themselves.

Much more careful transport experiments performed in later years showed that the situation was more complicated [10–16]. Already in 1991, Wang *et al.* [10] measured a positive Hall coefficient in these materials at low temperatures, and suggested that "*holes may, in fact, be the carriers responsible for superconductivity in* $Nd_{2-x}Ce_xCuO_{4-\delta}$." In 1993, Crusellas *et al.* [11] measured magnetoresistance and Hall effect and concluded that "*Holes dominate the low-temperature transport properties.*"

Even more detailed experiments followed. Figure 15.4 shows the titles of five articles and the year in which they were published. The first, by Tokura, Takagi, and Uchida already mentioned. The following three by Rick Greene, a researcher at University of Maryland (H-index 80) and coworkers [13–15]. Already in 1994, Greene informed that "*We find a remarkable correlation between the appearance of superconductivity and (1) a positive magnetoresistance in the normal state, (2) a positive contribution to the otherwise negative Hall coefficient, and (3) an anomalously large Nernst effect. These results strongly suggest that both holes and electrons participate in the charge transport for the superconducting phase of* $Nd_{2-x}Ce_xCuO_4$." In 1997, having performed more experiments, already the title of the paper suggested that there are "*two types of charge carriers*"

A superconducting copper oxide compound with electrons as the charge carriers (1989)

Anomalous Transport Properties in Superconducting $Nd_{1.85}Ce_{0.15}CuO_{4\pm\delta}$ (1994)

Thermomagnetic transport properties of $Nd_{1.85}Ce_{0.15}CuO_{4+\delta}$ films: Evidence for two types of charge carriers (1997)

Hole superconductivity in the electron-doped superconductor $Pr_{2-x}Ce_xCuO_4$ (2007)

Hole pocket–driven superconductivity and its universal features in the electron-doped cuprates (2019)

Fig. 15.4 Five articles published over the last 30 years on electron-doped cuprates. The year of publication is in parenthesis.

in these materials. Finally, in 2007 he concedes in the paper's title that there is "*Hole superconductivity in the electron-doped superconductor* $Pr_{2-x}Ce_xCuO_4$". The fifth article by Martin Greven [16], a researcher at University of Minnesotta (H-index 41) and coworkers, concludes in 2019 that their measurements together with those of previous authors definitively indicate "*a single underlying hole-related mechanism of superconductivity in the cuprates regardless of nominal carrier type.*"

We had anticipated all this. Already in 1989, together with my coworker Frank Marsiglio, we published an article [17] proposing an explanation for why when these compounds are doped with electrons, two types of charge carriers are created in two different bands, ones are electrons and the other holes, and that the holes determine the superconductivity. Precisely what these experimental researchers established over the next 30 years. But unfortunately, the majority of the scientific community continues to be confused about this. Even today, when I discuss with scientists the possibility that superconductivity may *need* holes, almost invariably the first reaction is: that can't be so, since there are cuprates where the charge carriers are electrons!

This is perhaps a good time to discuss where the electrons and the holes are in the cuprates.

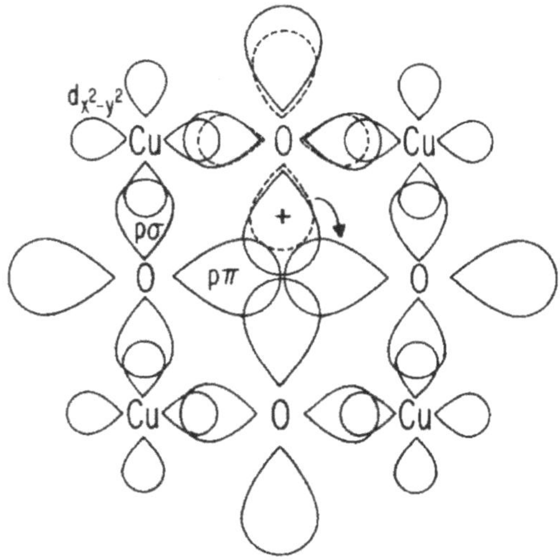

Fig. 15.5 Atomic orbitals in the copper–oxygen planes of the cuprates. The theory of hole superconductivity says that the charge carriers (holes) reside in the oxygen $p\pi$ orbitals, perpendicular to the Cu–O direction. In contrast, it is generally believed that the holes reside in the oxygen $p\sigma$ orbitals, hybridized with the copper $d_{x^2-y^2}$ orbitals.

Figure 15.5 shows part of Fig. 15.1, a copper–oxygen plane. The Cu has nominal valence Cu^{++}. This corresponds to having one hole in the orbital $d_{x^2-y^2}$, or equivalently one electron (each orbital can accommodate two electrons of opposite spin). Oxygen has nominal valence O^{--}, and all its p orbitals are full, with two electrons each. With that electronic content, the compounds are insulating. This is because the Coulomb repulsion prevents electrons from propagating through the copper $d_{x^2-y^2}$, even though those orbitals have a missing electron. This is what is called a "Mott insulator".

When the system is doped with holes, or equivalently when electrons are removed, there are two ways to do it, focusing on the oxygen orbitals. We can take electrons out of the orbital $p\sigma$, oriented in the direction of Cu, or from the orbital $p\pi$, oriented perpendicular to the Cu direction. This corresponds to two different bands in the system, the band $Op\sigma - Cud_{x^2-y^2}$ and the band $Op\pi$.

The universal consensus in the scientific community is that when the plane is doped with either holes or electrons, in both cases this occurs in the orbitals $Op\sigma$, that are hybridized with the Cu orbitals $d_{x^2-y^2}$. Electrons in orbitals $d_{x^2-y^2}$ have a strong repulsive interaction between them (called Hubbard U) and this interaction gives rise to magnetism. The consensus is that magnetism plays an important role in superconductivity.

Instead, for the last 30 years my collaborators and I propose that holes in the cuprates reside in the $Op\pi$ orbitals, and they give rise to superconductivity. We say that electrons or holes in orbitals $Op\sigma - Cud_{x^2-y^2}$, and the magnetism, is irrelevant. With respect to the electron-doped cuprates, we propose that in that case the doped electrons reside in $Op\sigma - Cud_{x^2-y^2}$ orbitals, and this induces holes in orbitals $Op\pi$, that in turn induce more electrons in orbitals $Op\sigma - Cud_{x^2-y^2}$. This explains naturally the Greene *et al.* experiments that see transport in two distinct bands, an electron band and a hole band. Instead, in the vision of the community both electrons and holes reside in orbitals $Op\sigma - Cud_{x^2-y^2}$ which is not easy to understand, since this is a single band. It requires a complicated elaboration taking into account electron-electron repulsion and translational symmetry breaking, which in my opinion is not valid.

But let us leave these technical details aside and continue with the story.

The attentive reader may have noticed an isolated peak in Fig. 15.2, that plots the yearly number of published articles on superconductivity with the word 'holes' in the title, in year 2001. There was a reason for that. In January 2001, the Japanese physicist Jun Akimitsu and collaborators published an article in the journal *Nature* titled "*Superconductivity at 39 K in magnesium diboride*" [18]. The article electrified the scientific community. It has 4,541 citations today, the yearly number of articles citing it is shown in Fig. 15.6.

The number of citations is impressive. But note that it was larger immediately after the compound MgB_2 was found, then it gets notably smaller. We will explain why.

Both elements Mg and B are relatively light and they don't have electrons in d orbitals, only in s and p. In general it is thought that

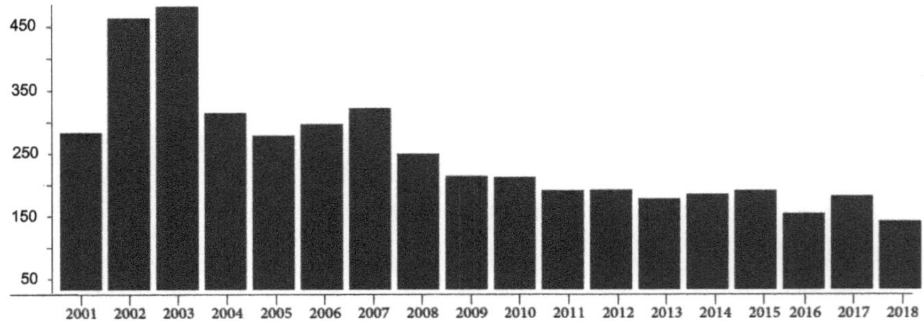

Fig. 15.6 Articles citing the discovery of superconductivity in magnesium diboride in 2001 [18].

metals with s and p electrons only are much simpler, electrons don't interact strongly between them, the Hubbard U is small. Thus it would be expected that MgB_2 is a relatively simple compound, and that if it is superconducting its superconductivity should be 'conventional', explained by BCS theory and the electron–phonon interaction. However, as we saw in earlier chapters it was expected that conventional superconductors couldn't have T_c's higher than about 25 K. The discovery of superconductivity at 39 K in MgB_2 was a great surprise to the community, and initially it generated confusion. For the scientific community, it was impossible that this superconductivity would have anything to do with the superconductivity of the cuprates, given that in MgB_2 there is no magnetism, there are no d electrons and there is no strong Hubbard interaction U as in the cuprates.

For me, it wasn't a surprise. On the contrary, I wanted to kick myself for not having thought about this compound earlier, since I could have *predicted* that it would be superconducting at high temperatures.

What MgB_2 has in common with the cuprates is precisely what I think is essential to have high temperature superconductivity: (1) Conduction by holes through overlapping p orbitals, (2) conduction through negatively charged ions, and (3) conduction in planes that are highly negatively charged. It doesn't have in common with the cuprates precisely what I believe has nothing to do with the superconductivity of the cuprates and others think it does: magnetism, d orbitals, and strong Hubbard U.

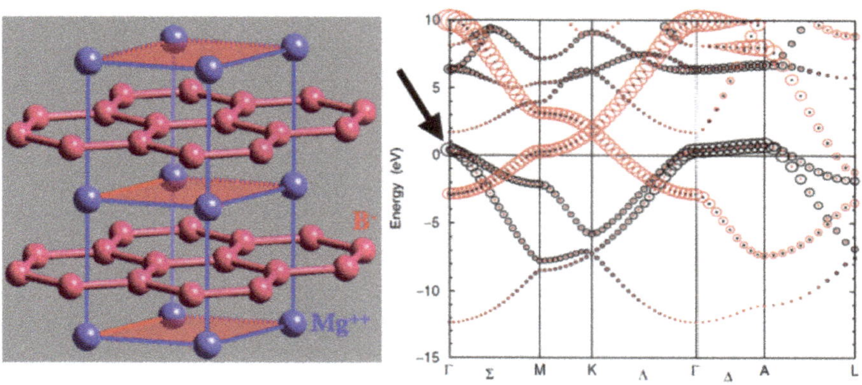

Fig. 15.7 Spatial structure (left) and band structure (right) of MgB_2 [19].

Figure 15.7 shows on the left the spatial structure of MgB_2. There are planes of ions B^-, separated by planes of ions Mg^{++}. It is known that the conduction that gives rise to superconductivity occurs through holes in the B^- planes. The right panel of Fig. 15.7 shows the band structure, calculated by Kortus *et al.* [19], that gives the energy of electronic states versus momentum. States with energy above zero are unoccupied. Where the arrow points there are two bands that cross zero almost at the Γ point. This indicates that those two bands are almost full, that is they have only a few holes. According to the Kortus paper, the radii of the red and black circles are proportional to the orbital character of the charge carriers in the boron orbitals: p_z (red) o p_x, p_y (black). The two almost full bands (black circles) are then bands formed out of p_x, p_y orbitals in the boron planes that connect directly, just like the $p\pi$ orbitals of oxygen in Fig. 15.5. The holes propagate through anions, i.e. negatively charged ions, B^- here, O^{--} in the case of the cuprates. That is precisely what leads to superconductivity within our theory, as we will see in Chapter 16. The planes formed by B^- ions are negatively charged, and so are the cuprate Cu–O planes, since the unit cell d $Cu^{++}(O^{--})_2$ has two excess negative charges.

Immediately after the announcement of the discovery of superconductivity in MgB_2, even before it had been experimentally determined that the charge carriers were holes, I wrote a paper titled "*Hole superconductivity in MgB_2: a high T_c cuprate without Cu*" where I pointed out that MgB_2 could be explained by the same hole mechanism that we had proposed already 12 years earlier for the cuprates. Initially my paper got some attention, it was cited 18 times in 2001. Shortly after my paper, Kang *et al.* [21] measured the Hall effect in MgB_2 and found that indeed the charge carriers were holes as I had anticipated. However, my explanation was rapidly eclipsed by the BCS professionals that launched a vigorous effort to explain how BCS and the electron–phonon interaction completely explain the 39 K superconductivity of MgB_2 [22, 23], even though nobody had predicted it.

Shortly after my paper, Warren Pickett proposed [22] that holes in boron p_x, p_y orbitals are strongly affected by vibrations of the boron, that have high frequency due to the light mass of boron, and this gives rise to a large electron–phonon interaction that according to him can explain the observed T_c within BCS theory. He comments that "*Although Hirsch has also focused on the hole character of the bands* [13], *his emphasis is otherwise quite different from that described here*". Certainly true. A few months later, Marvin Cohen and collaborators [23] pointed out that MgB_2 differs from ordinary metallic superconductors in various important aspects, they acknowledge that conventional models don't explain it and report the results of detailed calculations that they performed that supposedly show that the electron–phonon interaction is extremely anisotropic, varying between the limits 0.1 and 2.5. Taking into account this enormous anisotropy (a factor of 25), and in addition assuming that the phonons are anharmonic rather than harmonic, they inform that they can explain both the high transition temperature, 39 K, as well as the fact that the isotope coefficient is smaller than predicted by BCS (0.3 instead of 0.5).

A perfect example of the 'superflexibility' that Rainer was talking about.

The two articles that Cohen *et al.* wrote explaining MgB_2 [23] have more than 900 citations, Pickett's [22] has 674, many more than the 94 citations that my paper has [20] saying that the mechanism is hole superconductivity and is the same as for the cuprates. Clearly, they won.

To conclude this chapter, let us look again at Fig. 15.2. Again, in the number of yearly papers published with the word 'hole' in the title, there is a jump in the year 2008. Why?

In 2008, Hideo Hosono and coworkers (Japanese physicists) discovered a new class of superconductors, iron-arsenic compounds, or iron-selenium. Like the cuprates, they have planar structures. Like the cuprates, they have negatively charged anions (As^{---} or Se^{--}). Like the cuprates, the scientific community centers their attention on the cation (Cu^{++} in the cuprates, Fe^{+++} in the iron superconductors), I on the anions (O^{--}, As^{---}, Se^{--}). Like the cuprates, the majority of these compounds have holes as charge carriers, but some appear to have electrons. Like in the cuprates, we proposed that in these materials even when they are doped with electrons, hole carriers are generated [24]. The highest critical temperature for these compounds is about 60 K. The scientific community considers these compounds to be unconventional, not explained by BCS–electron–phonon, but there is no agreement on how to explain them, just like for the cuprates. Many believe that the mechanism in these compounds also has to do with magnetism, since they have antiferromagnetic states close to the doping regime where they become superconductors. However, they are not 'Mott insulators' in the undoped state, something that most physicists assume is essential to the superconductivity of the cuprates.

I believe that in these compounds, just like in the cuprates and all other superconductors, the mechanism has to do with holes and is always the same, as we will explain in the following chapters.

Returning to conventional materials, as we saw in Chapter 3 the great majority of superconducting elements have positive Hall coefficient indicating that holes dominate the conduction. Those that don't is because they have more than one type of carrier, that is they have both holes and electrons, and electrons have higher mobility in the normal state resulting in negative Hall coefficient. In other so-called conventional superconductors like the class of A15 materials, that have relatively high superconducting critical temperature, it has also been noted that the charge carriers are holes [25]. The same in the compound BaKBiO [26], which is similar to the cuprates but doesn't have Cu nor magnetism, and for which the critical temperature is relatively high, 30 K.

In Chapter 16, we will begin to understand why holes are necessary for superconductivity.

References

[1] David Lindley, Landmarks: Superconductivity explained, *Phys. Rev. Focus* **18**, 8 (2006).

[2] F. Marsiglio and J. P. Carbotte, Electron–Phonon Superconductivity", en Bennemann K.H., Ketterson J.B. (eds.) *Superconductivity.* Springer, Berlin, Heidelberg (2008).

[3] S. Uchida *et al.*, *Jpn. J. Appl. Phys.* **26**, L440 (1987).

[4] V. J. Emery, *Phys. Rev. Lett.* **58**, 2794 (1987).

[5] M. W. Shafer, T. Penney, and B. L. Olson, *Phys. Rev.* **36**, 4047 (1987).

[6] J. G. Bednorz and K. A. Müller, *Jpn. J. Appl. Phys.* **26**, Suppl. 26-3, 1781 (1987).

[7] Y. Tokura, H. Takagi, and S. Uchida, *Nature* **337**, 345 (1989).

[8] R. Pool, Update on electron superconductors, *Science* **243**, 1436 (1989).

[9] J. E. Hirsch, Hole superconductivity, *Phys. Lett. A* **134**, 451 (1989).

[10] Z. Z. Wang *et al.*, Positive Hall coefficient observed in single-crystal $\mathrm{Nd_{2-x}Ce_xCuO_{4-\delta}}$ at low temperatures, *Phys. Rev. B* **43**, 3020 (1991).

[11] M. A. Crusellas *et al.*, Two-band conduction in the normal state of a superconducting $Sm_{1.85}Ce_{0.15}CuO_4$ single crystal, *Physica C* **210**, I221 (1993).

[12] M. Suzuki *et al.*, Hall coefficient for oxygen-reduced $\mathrm{Nd_{2-x}Ce_xCuO_{4-\delta}}$, *Phys. Rev. B* **50**, 9434 (1994).

[13] Wu Jiang *et al.*, Anomalous transport properties in superconducting $\mathrm{Nd_{1.85}Ce_{0.15}CuO_{4\pm\delta}}$, *Phys. Rev. Lett.* **73**, 1291 (1994).

[14] P. Fournier *et al.*, Thermomagnetic transport properties of $\mathrm{Nd_{1.85}Ce_{0.15}}$ $\mathrm{CuO_{4+\delta}}$ films: Evidence for two types of charge carriers, *Phys. Rev. B* **56**, 14149 (1997).

[15] Y. Dagan and R. L. Greene, Hole superconductivity in the electron-doped superconductor $\mathrm{Pr_{2-x}Ce_xCuO_4}$, *Phys. Rev. B* **76**, 024506 (2007).

[16] Yangmu Li, W. Tabis, Y. Tang, G. Yu, J. Jaroszynski, N. Barisic, and M. Greven, Hole-pocket-driven superconductivity and its universal features in the electron-doped cuprates, *Science Advances* **5**, 7349 (2019).

[17] J. E. Hirsch and F. Marsiglio, On the dependence of superconducting T_c on carrier concentration, *Phys. Lett. A* **140**, 122 (1989).
[18] J. Nagamatsu *et al.*, *Nature* **410**, 63 (2001).
[19] J. Kortus, I. I. Mazin, K. D. Belashchenko, V. P. Antropov, and L. L. Boyer, *Phys. Rev. Lett.* **86**, 4656 (2001).
[20] J. E. Hirsch, Hole superconductivity in MgB_2: A high T_c cuprate without Cu, *Phys. Lett. A* **282**, 392 (2001).
[21] W. N. Kang *et al.*, Hole carrier in MgB_2 characterized by Hall measurements, *Appl. Phys. Lett.* **79**, 982 (2001.
[22] J. M. An and W. E. Pickett, *Phys. Rev. Lett.* **86**, 4366 (2001).
[23] H. J. Choi *et al.*, *Phys. Rev. B* **66**, 020513(R) (2002); *Nature* **418**, 758 (2002).
[24] F. Marsiglio and J. E. Hirsch, *Physica C* **468**, 1047 (2008).
[25] L. Hoffmann, A. K. Singh, H. Takei, and N. Toyota, *J. Phys. F: Met. Phys.* **18**, 2605 (1988).
[26] J. H. Lee *et al.*, *Phys. Rev. B* **61**, 14815 (2000).

Part IV

ELECTRON–HOLE ASYMMETRY AND ITS CONSEQUENCES

Chapter 16

The fundamental asymmetry between electrons and holes

In Fig. 16.1, that I have had for many years on my website, I give an intuitive idea of some differences between electrons and holes. We will see where these differences come from, and what they have to do with superconductivity. We will also try to understand why in the scientific community it is erroneously believed that electrons and holes are by and large equivalent.

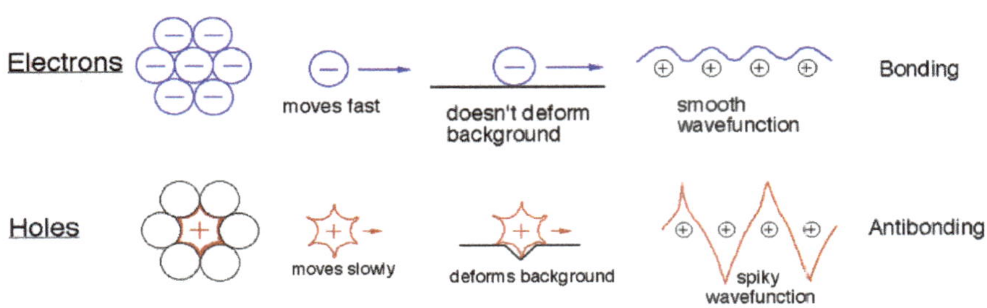

Fig. 16.1 The hole is the absence of an electron. As can be seen in this figure, holes are very different from electrons.

In a solid, electronic states are organized in bands. Figure 16.2 shows a very simple band.

The vertical axis in this figure is the energy of the electron, the horizontal axis is what is called the crystal momentum, that represents the inverse of the wavelength of the electron. Recall that in quantum mechanics, particles are also waves, and waves have

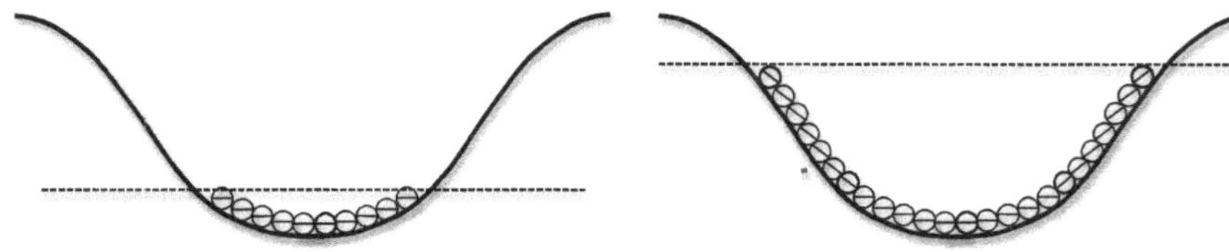

Fig. 16.2 Band of electronic states. Energy as function of crystal momentum. The dashed line indicates the highest energy of occupied states, called Fermi energy, or Fermi level.

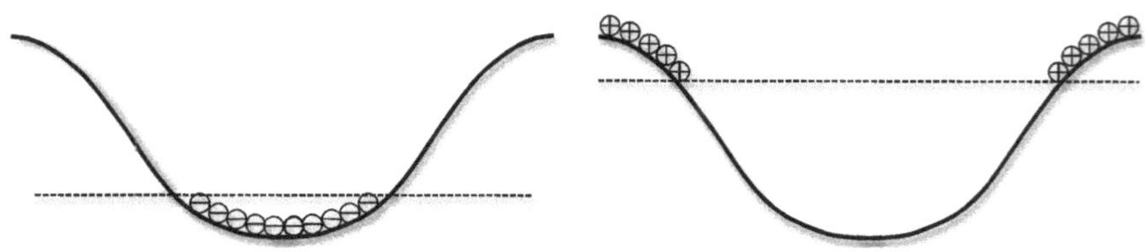

Fig. 16.3 Like Fig. 16.2, but on the right panel we show the missing electrons in the band, the holes.

wavelengths. The dashed line indicates the highest energy of occupied states, called the Fermi level or Fermi energy.

The left panel shows a band that is almost empty, in that case, we call the charge carriers electrons. The right panel shows a band that is almost full, in that case, we call the charge carriers holes. If we measure the Hall effect, we will find a negative value for a metal where the conduction band is close to empty as in the left panel, and a positive value for a metal where the conduction band is close to full, as in the right panel. Equivalently, we can think that the charge carriers on the right are holes, of positive charge, as Fig. 16.3 shows.

Bands can be also oriented in the reverse way, with the energy maximum at the center of the band rather than at the edges, as Fig. 16.4 shows. Here, the holes are near the center of the band, what is called the Γ point, as was seen in Fig. 15.7.

The electrical conductivity depends on the number of carriers and the filling of the band. If the band is either totally empty or totally full, it does not conduct electricity. In the case of Figs. 16.3 and 16.4, the conductivity is proportional to the number of electrons for the

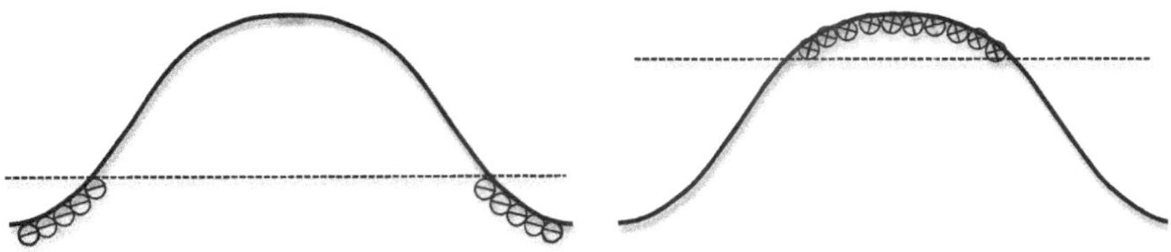

Fig. 16.4 Bands can also be oriented such that the maximum is at the center.

left panel and to the number of holes for the right panel. For that reason, it is more convenient to think about the holes rather than to think about all the electrons on the right panel of Fig. 16.2, the latter are much more numerous but they don't contribute to the electrical conductivity, in fact, those that are above the middle of the band 'anticontribute'.

For many purposes, like the one I just explained, one can think about holes as equivalent to electrons. One can make a mathematical transformation (called a 'particle–hole transformation') that transforms the right panel of Fig. 16.3 or Fig. 16.4 into the left panel, changing also the sign of the charge. Physicists like mathematics and that led them, starting with Heisenberg in 1931, to conclude that 'electron–hole symmetry' is something very common, practically universal. That is, an almost full band behaves the same as an almost empty band, except for the fact that the sign of the Hall coefficient is opposite. But it is not so. The apparent symmetry in Fig. 16.3 or Fig. 16.4 between electrons and holes hides a profound asymmetry that exists in nature between electrons and holes that is in part revealed in an intuitive way with what Fig. 16.1 shows. Let's see.

The left panel of Fig. 16.5 shows qualitatively electronic states in a solid, the wavefunction of an electron. When the electron is close to the bottom of the band, the wavefunction (called 'bonding') is as shown in the lower part, when the electron is near the top of the band, the wavefunction (called 'antibonding') is as shown in the upper part. Note that the wavefunction in the lower part is never zero and varies more or less smoothly with larger amplitude where the positive ions are and smaller in the interstitial region between ions. This makes sense because it minimizes both the potential energy and the kinetic

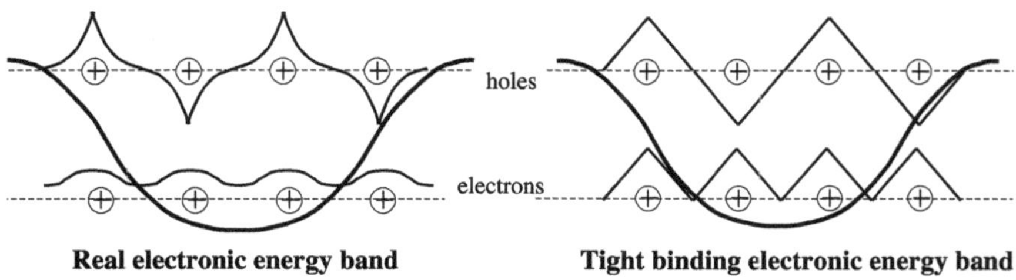

Fig. 16.5 Left panel: wavefunction of electrons near the bottom of the band "bonding" and near the top of the band "antibonding". The circles with + signs denote the ions. Right panel. Approximation used in mathematical models that do not reflect reality.

energy of the electron. The quantum kinetic energy increases when the wavefunction oscillates more.

Now, the Pauli exclusion principle states that each state in the band can be occupied by no more than two electrons. As we add electrons to the band, we need to occupy states of increasingly higher energy. Near the top of the band, the states have wavefunction as shown in the upper left panel, called "antibonding", this is also the wavefunction of the 'holes' when the Fermi level is close to the top of the band, which is when we use the hole concept. Here, the wavefunction has many more oscillations, reflecting the higher kinetic energy, and is zero at the midpoint between two ions.

The electronic charge density is proportional to the square of the wavefunction. This tells us that the "bonding" states have higher charge density between the ions, this effectively causes the ions to attract each other and gives stability to the solid, that's why they are called "bonding". The "antibonding" states have zero (or low) charge density between the ions, so there the repulsive interaction between ions predominates and the ions repel each other, that's why they are called "antibonding" and they make the solid unstable. If there are too many "antibonding" electrons, the solid will make a phase transformation to another structure that allows it to have fewer antibonding states occupied.

The right panel of Fig. 16.5 shows a 'caricature' of the left panel. In it, the wavefunction of the antibonding electron, or of the hole, is practically equivalent to that of the electron in the sense that one can make a mathematical transformation that changes the sign of the wavefunction with periodicity twice the periodicity of the ionic lattice, and the wavefunction of holes becomes the wavefunction of electrons. This caricature no longer reflects the physical difference between bonding and antibonding states, it has electron–hole symmetry. In the models that physicists use to describe superconductivity and other phenomena, for example, the Hubbard model, the wavefunctions are assumed to be as shown in the right panel, and one of the fundamental physical differences between electrons and holes is lost.

As we said, the occupied antibonding states generate instability in the structure of the solid. The antibonding states are occupied when bands are almost full, which is the case where the charge carriers are holes. So, we would expect that if holes are important for superconductivity, there should be a correlation between the existence of superconductivity and instabilities in the solid. This was in fact noticed by Bernd Matthias in his extensive investigations trying to find new superconductors [1]. He found that when superconductivity became stronger, the critical temperature increased, it was associated with the solid becoming more unstable, and there came a point where instead of T_c continuing to rise, the solid changed its structure to another more stable one where superconductivity disappeared. In the new more stable structure, there probably were no antibonding electrons, hence there were no holes. This observation by Matthias supports the hypothesis that holes are important for superconductivity.

But, to be fair, we should mention that BCS has an alternative explanation for the correlation found by Matthias between lattice instabilities and superconductivity — that when the electron–phonon interaction is large, it gives rise both to a high T_c as well as to a tendency to instability in the lattice structure.

Another fundamental difference between electrons and holes results from the Coulomb interaction between electrons. Let's see how to understand that.

Figure 16.6 shows schematically a hydrogen atom with zero, one, and two electrons, that is, the ions H^+, H, and H^-. The atomic orbital is described by a wavefunction $\varphi(r)$. This figure has electron–hole

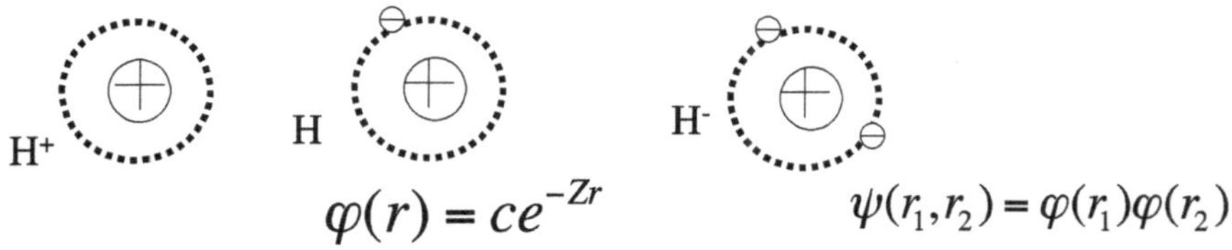

Fig. 16.6 Hydrogen atom with zero, one and two electrons in the approximation used in the Hubbard model. The wavefunction is given in atomic units, where r is in units of a_0, the Bohr radius.

symmetry, and this is how the Hubbard model would describe the hydrogen atom. The Hubbard model says that when there are two electrons in the orbital (the ion H^-), there is a repulsive interaction energy, the famous Hubbard U. But it also says that the wavefunction of the two electrons is simply the product of two of the same wavefunctions for the one electron in the H atom. And that is wrong. Figure 16.7 shows reality, or something much closer to reality.

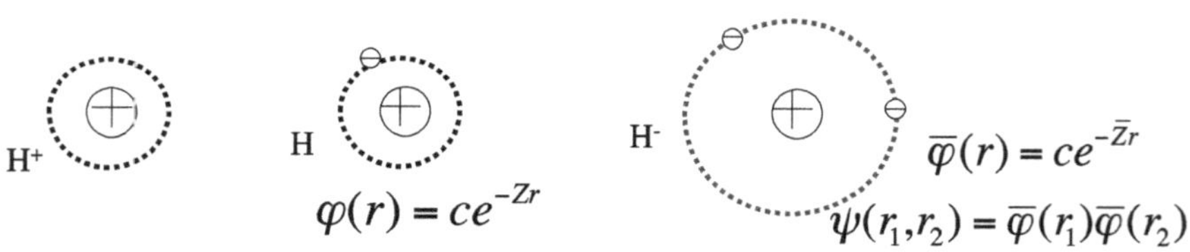

Fig. 16.7 Hydrogen atom with zero, one, and two electrons taking into account the orbital expansion caused by Coulomb repulsion between electrons.

When there are two electrons in the orbital, the repulsive interaction modifies the wavefunction. The wavefunction expands to reduce the repulsive interaction between electrons. That costs electron–ion interaction energy that is attractive. But it has another benefit — to reduce the quantum kinetic energy of the electrons. Balancing these three energies, one finds the optimal expansion that minimizes the total energy. The answer is that the modified wavefunction has in the exponent the quantity $\bar{Z} = Z - 5/16$ instead of the quantity Z that the wavefunction for one electron has.

What is the meaning of Z and $\bar{Z}$? Z is the atomic number of the atom, $Z = 1$ in this case. But we call it Z so we can generalize. If instead of hydrogen ions H^+, H, H^- we consider the helium ions

He^{++}, He^{+}, He, also with zero, one and two electrons, we have the same wavefunctions as in the figure with $Z = 2$. The wavefunctions are given in atomic units, where the unit of length is the Bohr radius $a_0 = 0.529\mathring{A}$. The radius of the wavefunction is approximately a_0/Z for one electron and $a_0/\bar{Z}$ for two electrons. Since $\bar{Z} < Z$, the radius expands. In Fig. 16.7, there no longer is electron–hole symmetry. Another way to say this is that Z decreases to $\bar{Z}$ because one electron partially screens the ionic charge that the other electron sees.

In the band that is generated when we put hydrogen ions in a lattice to form a crystal, if the band is almost empty almost all the atoms are H^{+}, and a few are H, with radius a_0/Z. Instead, if the band is almost full, then almost all the ions are H^{-} with the expanded orbits of radius $a_0/\bar{Z}$. This is shown schematically in Fig. 16.8. Clearly, there is no electron–hole symmetry.

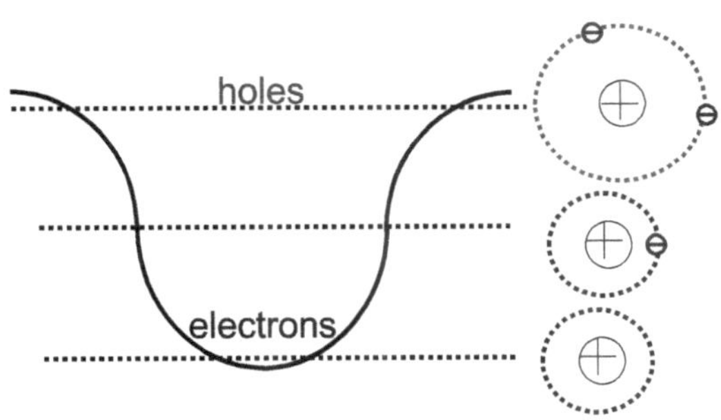

Fig. 16.8 Band in a solid composed of hydrogen atoms. Close to the bottom of the band, atoms occupied by electrons have a single electron and the radius of the orbit is not expanded. Close to the top of the band, almost all the atoms have two electrons with the orbits expanded.

Let us next consider the motion of charge carriers when the band is almost empty and when it is almost full. This is shown schematically in Fig. 16.9. The lower panel shows the propagation of an electron in the almost empty band. It is very simple, the electron simply hops from an atom to its nearest neighbor atom that has no electron, nothing else happens. Instead, the propagation of holes in the almost full band, shown in the upper panel, is very complicated. When the electron jumps from the ion on the left H^{-} to the atom H at the center, two other things happen: the other electron on the

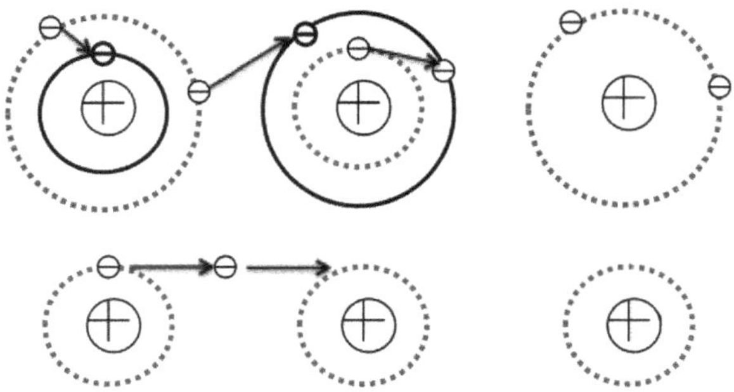

Fig. 16.9 Propagation of electrons (lower panel) and of holes (upper panel) in a solid composed of hydrogen atoms. The lower (upper) panel shows what happens when the band is almost empty (full).

left atom reduces its orbit, and the other electron on the center atom expands its orbit.

In other words, there is a large *asymmetry* in how charge propagates in an almost empty and an almost full band, or equivalently there is a large asymmetry in how charge propagates when carriers are electrons and when they are holes. This asymmetry is lost if we forget that the orbital expands when it is doubly occupied, as the Hubbard model assumes.

We may say that the propagation of the hole generates a disruption in its environment, the propagation of the electron does not. This causes the hole to have greater difficulty to propagate. We may think of the electron propagating as if it was ice skating, the hole propagating as if it was wading through a swamp. Figure 16.1 shows a caricature of this, the hole 'deforms' its environment when it propagates, the electron does not. One consequence of this is that the hole has an "effective mass" m^* that is bigger than that of the electron: the hole is heavy, the electron is light.

What does this have to do with superconductivity? To understand it qualitatively, let us consider Fig. 16.10.

Where would the reader prefer to park his/her car? In an almost empty garage or an almost full one? To move the car in an almost empty garage (a) is very easy. To move a 'car hole' in an almost full garage (b) is much more complicated, we have to move

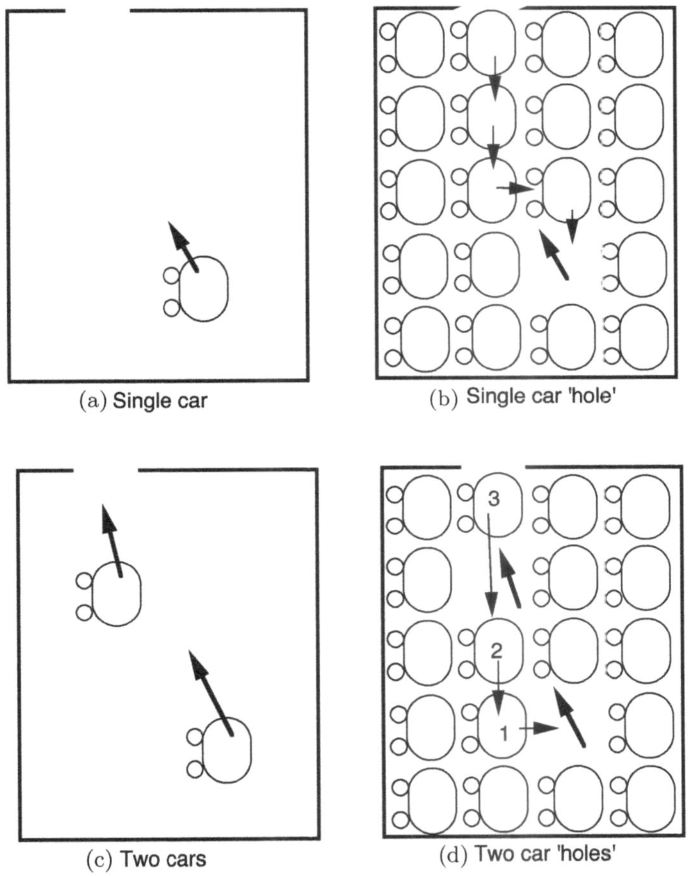

Fig. 16.10 Movement of cars in a garage that is almost empty ((a), (c)) and of 'car holes' in a garage that is almost full ((b), (d)).

many other cars, there is risk of collisions, etc. To move two cars (c) is also easy, and it is convenient to keep them far from each other to avoid collisions. Instead, to move two 'car holes' it is convenient to follow the sequence shown in (d) 1, 2, 3. After 1 and 2, the holes are 'paired' and it is easier to move them.

As BCS taught us, superconductivity requires that charge carriers pair. Electron–hole asymmetry makes it easier for holes to propagate if they pair than if they don't pair. Propagation is good, the wavefunction expands and in quantum mechanics expansion of the

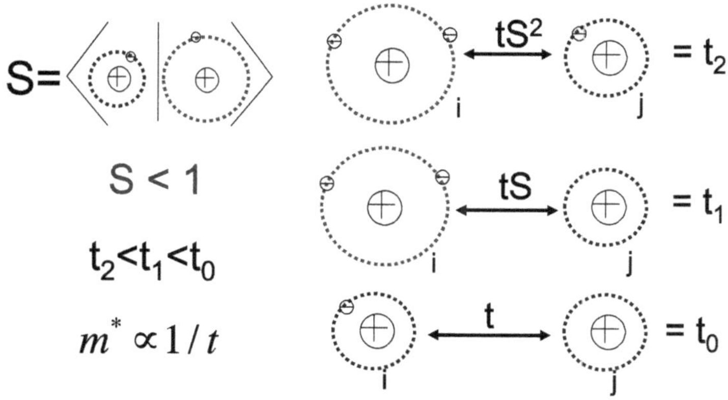

Fig. 16.11 The parameter $S < 1$ measures how much the expanded orbital differs from the unexpanded one. The more it differs, the smaller S is. The hopping amplitude for an electron to a neighboring site depends on the occupation of the sites, it gets smaller for higher occupation.

wavefunction lowers the kinetic energy. So, when the band is almost full and there are a few holes, holes will pair to propagate more easily and that way lower their kinetic energy, and the pairing gives rise to the superconducting state.

More quantitatively, Fig. 16.11 shows the hopping amplitude for an electron to neighboring atoms. Because of the expansion of the orbital when it is doubly occupied, the hopping amplitude t gets multiplied by factors $S < 1$ as the occupation in the atoms increases. The 'effective mass' of a charge carrier is proportional to the inverse of the hopping amplitude, so for a hole the effective mass is $m_2^* \propto 1/t_2$ and for an electron it is $m_0^* \propto 1/t_0$, so $m_2^* > m_0^*$, the hole is heavier than the electron.

Figure 16.12 shows what happens in a band. We can think about it equivalently in hole language or in electron language, looking at the right or the left side of the figure. Recall that hole is simply absence of electron. Figure 16.12 shows that as the number of electrons in the band increases, the hopping amplitude of electrons at the Fermi energy (dotted lines) decreases, since $t_0 > t_1 > t_2$. Equivalently, as the number of holes in the band increases, the hopping amplitude for holes at the Fermi level increases.

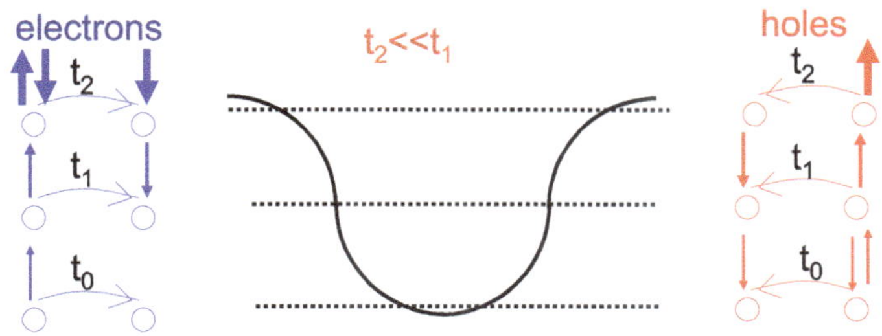

Fig. 16.12 Hopping amplitudes for electrons or holes at the Fermi energy, as function of position of the Fermi level in the band (dotted line). Hopping amplitude is maximum at the bottom of the band and becomes gradually smaller toward the top of the band.

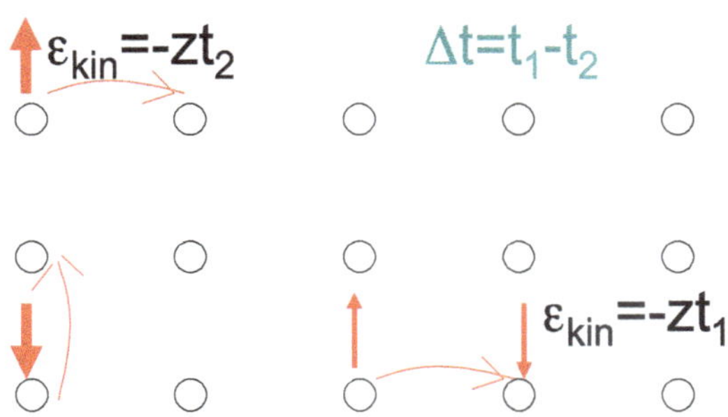

Fig. 16.13 Kinetic energy ϵ_{kin} of isolated holes and of paired holes. The kinetic energy becomes lower when holes pair.

The quantity that is important for superconductivity is the difference $\Delta t = t_1 - t_2$ that causes holes to pair. It would be the equivalent of the electron–phonon interaction λ in BCS theory. As Fig. 16.13 shows, when a hole hops from one site to the next, its kinetic energy is $\epsilon_{\text{kin}} = -zt_2$, where z is the number of nearest neighbor sites (for example, $z = 4$ for a square lattice). When a hole lands on a site where there is already another hole (of opposite spin), its kinetic energy is $\epsilon_{\text{kin}} = -zt_1$. Note that these energies are negative. Because $t_1 > t_2$, holes lower their kinetic energy when they hop to sites where there is another hole, and this makes them want to pair to lower their energy. As BCS taught us, pairing leads to superconductivity.

This pairing mechanism [2] is only effective when the Fermi level is close to the top of the band, that is, the band is almost full. Just like in the garage of Fig. 16.10, it is advantageous to pair the car holes when the garage is almost full of cars. As the garage becomes emptier, it becomes indifferent whether to pair or not. Recall also that holes, just as electrons, repel each other due to the Coulomb repulsion between charges of the same sign. So, this 'kinetic' pairing mechanism competes with the repulsive interaction, and only wins when it is most effective, close to the top of the band. The Coulomb interaction does not change with band occupation.

Incidentally, pairing through this interaction Δt leads naturally to the existence of a positive isotope effect. Δt increases somewhat on average when the mass of the ion goes down because the zero point quantum fluctuations of the ions increase, and this leads to a small increase in T_c. This has nothing to do with the electron–phonon mechanism of BCS, but offers an alternative explanation for the experimental observations of isotope effect.

Using equations from BCS theory but with this interaction instead of the electron–phonon interaction, we obtain the critical temperature versus hole concentration in the band shown in the left panel of Fig. 16.14. T_c first increases when the number of holes increases, reaches a maximum and then decays to zero when there are too many holes in the band.

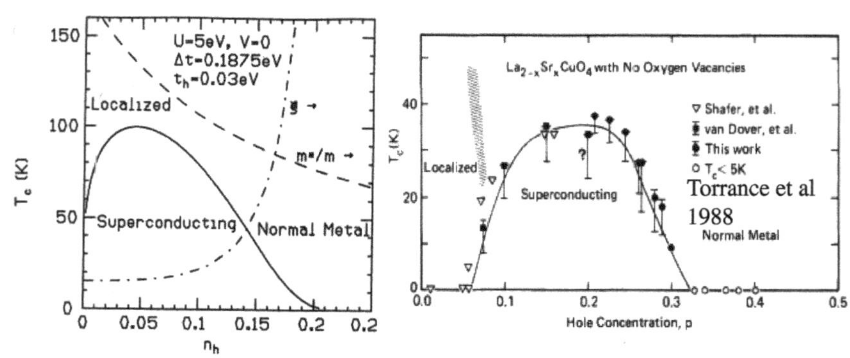

Fig. 16.14 Left panel: results of calculations of critical temperature versus hole concentration n_h per oxygen using the model of hole hopping explained in the text. The right panel shows the critical temperature versus hole concentration per unit cell in a cuprate superconductor according to experiments by Torrance *et al.* [3]. The relation between the horizontal scales is $p = 2n_h$.

The right panel of Fig. 16.14 shows the behavior of T_c versus hole density for a cuprate [3]. The behavior is similar for other cuprates and qualitatively similar to what we obtain from the theoretical calculation.

This asymmetry between electrons and holes has consequences that can be measured experimentally. In collaboration with Frank Marsiglio, who was a postdoctoral research associate at UCSD at that time, we investigated in detail the properties of the hole model with modulated hopping of Fig. 16.12 [2]. An interesting consequence that we found was that in tunneling experiments like the ones we discussed in Chapter 14 in connection with BCS, there should be an asymmetry with respect to the sign of the voltage across the junction: the current should be larger when the superconductor is negatively biased, that is, when the superconductor is ejecting electrons. When we did this calculation, there was no experimental evidence for this, but several years later it was confirmed that this is so for the cuprates. Figure 16.15 shows results of the theoretical calculation done in 1989 [2], and experimental measurements done in 1998 [4].

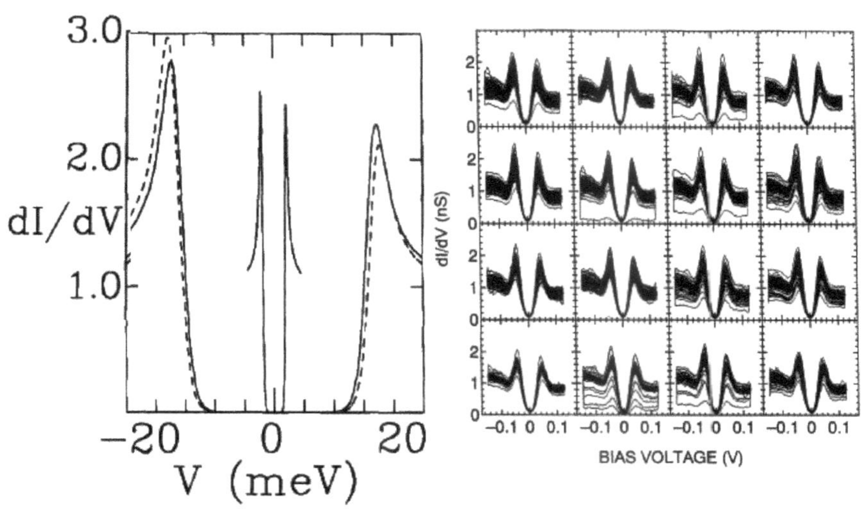

Fig. 16.15 Tunneling conductance for the model of hole superconductivity discussed in the text (1989) [2] (left panel) and experimental measurements done in 1998 in cuprates [4] (right panel).

It can be seen that the experiment confirms the prediction of the theoretical calculations done 9 years earlier. Other tunneling experiments show similar results [5].

The interaction that gives rise to superconductivity in this model is $\Delta t = tS(1 - S)$, where S measures the overlap of the expanded and unexpanded orbital, Fig. 16.11. If the orbital expands very little, S is close to 1 and Δt is small. The expansion of the orbital depends on the effective charge of the ion, Z, and is the largest when Z is the smallest, that is, when the ion attracts the electrons less strongly. For example, Δt is larger for a solid of ions H^- with a few holes than for a solid of atoms He^0 with a few holes. This is why the most favorable condition for superconductivity is when holes propagate through a lattice of negative ions, for example, B^- in MgB_2 or O^{--} in the cuprates. The more negative the ions are, the better.

Going back to the dependence of T_c on the number of holes, a behavior similar to that seen in the right panel of Fig. 16.14 is also found in transition metal alloys, investigated by Matthias and others in the 50s and 60s, that gave rise to the famous 'Matthias' rules' to describe the behavior of T_c versus number of electrons per atom. These were empirical rules that were never explained.

Figure 16.16 shows those measurements. It can be seen that there are two 'domes' as function of the average number of electrons per

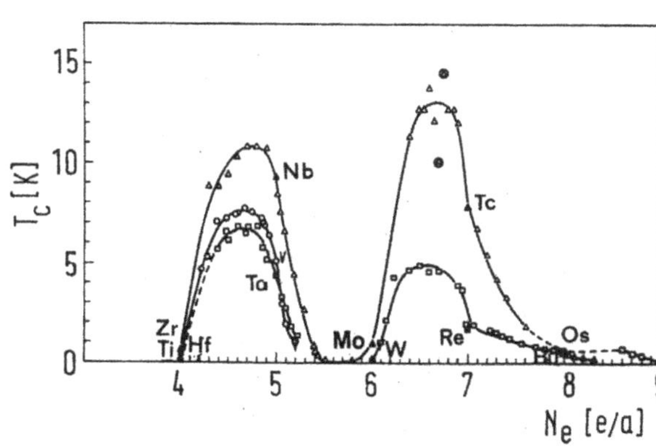

Fig. 16.16 Critical temperature in transition metal alloys, as function of the average number of valence electrons per atom e/a [6].

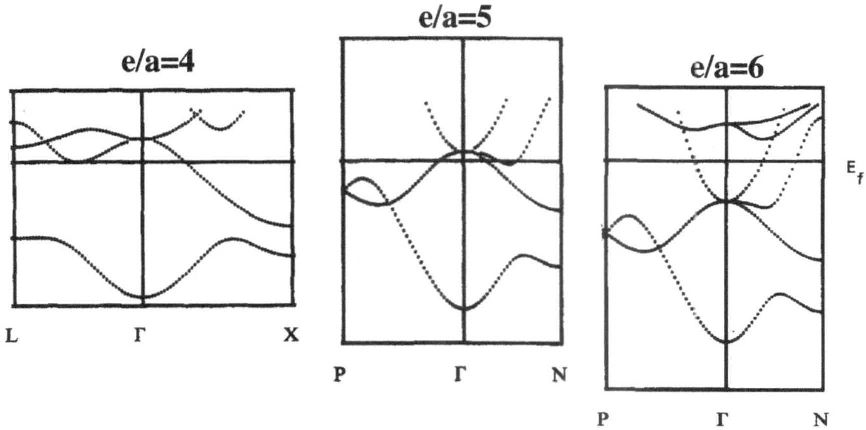

Fig. 16.17 Band structure of the metals Zr, Nb, and Mo with valence electrons per atom e/a = 4, 5, and 6.

atom, e/a. These domes are qualitatively similar to the behavior of T_c in the cuprates, right panel of Fig. 16.14. As in the cuprates, we can correlate this behavior with the hole occupation in the bands. Here, it is more complicated because these are three-dimensional systems and several bands play a role. For example, looking at the band structure of the metals Zr, Nb and Mo (Fig. 16.17), it can be simply understood why T_c decreases after V and Mb (e/a = 5) to zero between e/a = 5 and e/a = 6. Starting from the central panel of Fig. 16.17 with e/a = 5, adding more electrons, the Fermi level goes beyond the top of the band at the Γ point precisely where T_c goes to zero, around e/a = 5.5. Calculations with the model of hole superconductivity reproduce this behavior [7].

In summary, electrons and holes in electronic energy bands are very different. Electron–hole symmetry is a myth. Models that don't take into account electron–hole asymmetry, like the Hubbard model, don't reflect the real world. Electron–hole asymmetry is a necessary consequence of the fact that electrons and protons have the same magnitude of charge with opposite sign, but masses that differ by a factor of 2000. In the following chapters, we will understand more consequences of this asymmetry.

References

[1] B. T. Matthias, *Physica* **69**, 54 (1973).

[2] J. E. Hirsch and F. Marsiglio, *Phys. Rev. B* **39**, 11515 (1989); *Physica C* **162–164**, 591 (1989).

[3] J. B. Torrance *et al.*, Anomalous disappearance of high-Tc superconductivity at high hole concentration in metallic $La_{2-x}Sr_xCuO_4$, *Phys. Rev. Lett.* **61**, 1127 (1988).

[4] S. Kaneko, N. Nishida, K. Mochiku, and K. Kadowaki, *Physica C* **198**, 105 (1998).

[5] Ch. Renner *et al.*, *Phys. Rev. Lett.* **80**, 149 (1998).

[6] S. V. Vonsovsky, Y. A. Izyumov, and E. Z. Kurmaev, *Superconductivity of Transition Metals*, Springer, Berlin, 1982.

[7] X. Q. Hong and J. E. Hirsch, Superconductivity in the transition metal series, *Phys. Rev. B* **46**, 14702 (1992).

Chapter 17

Superconductors as 'giant atoms'

The idea that superconductors are in a way 'giant atoms' was proposed by many early researchers in superconductivity, as we saw in Chapter 11. However, they were only referring to their magnetic properties, not to their charge distribution.

In 2001, 12 years after starting to develop the theory of hole superconductivity, I came to the conclusion that when a metal enters the superconducting state, it expels electrons from the interior to the surface. Figure 17.1 is from a paper I wrote in 2001 [1]. It looks indeed like a 'giant atom'.

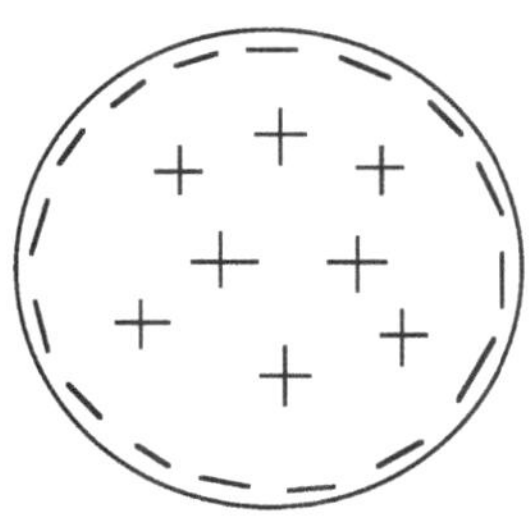

Fig. 17.1 Schematic image of a superconducting spherical body. Negative charge is expelled from the interior to the surface.

The first natural question is, how is it possible that in a material that conducts electricity a charge configuration like that shown in Fig. 17.1 can exist? A normal metal certainly cannot have that charge configuration because it gives rise to an electric field in the interior. But a metal cannot have an electric field in its interior (if it is not conducting electricity). If it did, mobile charges in the metal would move propelled by the electric force to positions that nullify the electric field, which gives the state of lowest energy.

For that reason, the possibility that superconductors may have electric fields in their interior had never been considered before. However, the argument just given for metals does not apply to superconductors. The reason is, the superconductor is in a macroscopic quantum state. We can think that the entire set of electrons in the superconductor, the so-called 'superfluid', is described by a macroscopic wavefunction $\Psi(\vec{r})$, entirely analogous to the wavefunction $\varphi(\vec{r})$ that describes a single electron in the hydrogen atom. The charge distribution in the hydrogen atom certainly is not homogeneous. Why not? Because it minimizes the sum of kinetic energy and potential energy. The potential energy wants to bring the negative electron as close to the positive nucleus as possible, but that increases the kinetic energy. The kinetic energy wants to get the electron as far from the nucleus as possible, but that increases the potential energy. The result is a compromise, the Bohr radius a_0.

Similarly, in the superconductor there is no *a priori* reason for why the distribution of negative charge has to be macroscopically homogeneous as in a normal metal. That minimizes the potential energy only, which is all that matters at the macroscopic level for the normal metal because the normal metal is not in a macroscopic quantum state. But the superconductor is.

Quantum physics tells us that when the wavefunction expands, the quantum kinetic energy is lowered. An expression for the quantum kinetic energy is

$$E_{\text{kin}} = \frac{\hbar^2}{2mr^2} \tag{17.1}$$

where $\hbar$ is Planck's constant, m is the particle mass, and r is the radial extension of the wavefunction. The wavefunction wants to expand (increase r) to lower its quantum kinetic energy. We can call that 'quantum pressure'. According to Eq. (17.1), quantum pressure is larger the smaller m is, i.e. the lighter the particle is.

If the readers ask themselves where Eq. (17.1) comes from, we can give an explanation based on the well-known Heisenberg uncertainty principle (the same Heisenberg of the holes). I am not satisfied with that explanation, I think there is a better one, but let's leave that for later. According to Heisenberg, the uncertainty in the position Δx and the uncertainty in the momentum Δp satisfy the condition

$$\Delta x \Delta p \sim \hbar. \tag{17.2}$$

The kinetic energy in classical mechanics is given by the expression

$$E_{\text{kin}} = \frac{1}{2}mv^2 = \frac{p^2}{2m} \tag{17.3}$$

where $p = mv$ is the momentum and v is the velocity. We may say that the uncertainty in the position Δx is of order r. And, assuming that on average p is zero, $\langle p \rangle = 0$, we have $(\Delta p)^2 = \langle p^2 \rangle - \langle p \rangle^2 = \langle p^2 \rangle$, and we deduce that on average Eq. (17.3), using the condition Eq. (17.2), gives Eq. (17.1).

And that is the essence of the charge asymmetry that leads to electron–hole asymmetry and to superconductivity. Ions are heavy, electrons are light. Electrons have much more quantum kinetic energy than ions, and they exert much more quantum pressure than ions, so they will expand their wavefunction more than ions, and the result is that in a macroscopic quantum system with negative electrons and positive ions naturally there will be more negative charge near the surface, and by charge neutrality more positive charge in the interior. The tunneling asymmetry shown in Fig. 16.15 is also a manifestation of this tendency of superconductors to expand the electronic wavefunction, which leads to outward radial motion of negative charge and hence to negative charge expulsion: as Fig. 16.15 shows, the conductance and for that reason the electric current are larger when the superconductor is negatively biased, that is, when electrons are coming out of the superconductor.

This tendency of superconductors to expel negative charge is naturally understood within the theory of hole superconductivity. Metals become superconductors only when there are many electrons in the band, the band is almost full. In addition, the kinetic energy of the charge carriers is very high, as shown in Fig. 16.5. It is natural that the material expels electrons when it has many, and it is natural that the mechanism is related to lowering of kinetic energy when the initial kinetic energy is high. And the charge is expelled when the wavefunction expands, and this lowers the kinetic energy according to Eq. (17.1). Besides, as we said earlier, the tendency to superconductivity is larger when the effective valence of the ion Z is smaller, which corresponds to having less positive charge, hence more excess negative charge in the system. It is natural that the system wants to get rid of the excess negative charge.

The next question is, how large is this charge inhomogeneity?

We need to take into account that charge inhomogeneity costs potential energy. Where does that energy come from? If this is associated with superconductivity, it is natural that this energy comes from the condensation energy of the superconductor, i.e. the difference in energy between the normal and superconducting states. And that this condensation energy should be quantum kinetic energy. The potential energy density in the presence of an electric field E is given according to the laws of electromagnetism by the expression

$$\frac{E^2}{8\pi}. \tag{17.4}$$

For a given quantity of charge expelled, we can calculate the resulting electric field in the interior, E, and, integrating over the volume of the body, we obtain the total cost in electrostatic energy. Equating this to the condensation energy, we can estimate the amount of charge expelled. We will leave it for later. In order of magnitude, near the surface, there is approximately one excess electron for every one million ions.

It is notable that in the early works on superconductivity where it was said that superconductors were like 'giant atoms', this was said with reference to their magnetic properties, and nobody thought of thinking that a 'giant atom' naturally should have inhomogeneous charge distribution, with more negative charge outside and more positive charge inside, as the microscopic atoms have.

In the first paper where I proposed this [1], I explained how I reached this conclusion based on an analysis of the solution of the model discussed in the previous chapter. But at that time I didn't yet understand in detail what the charge distribution would be like. I only guessed [1] that the excess negative charge would be located within the region of thickness λ_L, the London penetration depth, next to the surface because that is the region where the supercurrent flows. In later work, I confirmed that this is so, as we will see later.

The idea that the superfluid is described by a wavefunction $\Psi(\vec{r})$ completely analogous to the wavefunction of a single electron in the hydrogen atom is implicit in BCS theory, but more explicitly was articulated by two Russian physicists, Ginzburg and Landau (GL), in 1950, 7 years before BCS. After BCS, the Russian physicist Gorkov showed in 1958 that BCS implies GL. A few years later, the English physicist Brian Josephson explained how the quantum phase of the wavefunction $\Psi(\vec{r})$ can be measured experimentally in tunneling experiments between two superconductors. For that reason, the phenomenon is called 'Josephson tunneling', and it earned him the Nobel prize in 1973, one year after BCS's, together with Ivan Giaver who had measured the density of states in superconductors using tunneling between a superconductor and a normal metal, as we saw in Chapter 14.

Notably, none of these works considered the *electric charge* associated with the wavefunction $\Psi(\vec{r})$, in particular its sign. That is, in BCS–GL theory, the charge of $\Psi(\vec{r})$ can be negative or positive or zero. Instead, in the theory of hole superconductivity, $\Psi(\vec{r})$ necessarily has negative charge. We have shown theoretically that it is not possible to have Josephson tunneling between superconductors with opposite charge of $\Psi(\vec{r})$, i.e. one negative, one positive [2]. The fact that Josephson tunneling has been found to exist between practically all superconducting materials implies that $\Psi(\vec{r})$ has to have the same unique sign for the charge in all superconductors, as the theory of hole superconductivity predicts, contrary to BCS.

Another consequence of the fact that superconductors are like 'giant atoms' with macroscopically inhomogeneous charge distribution, as shown in Fig. 17.2, is that there will be macroscopic currents

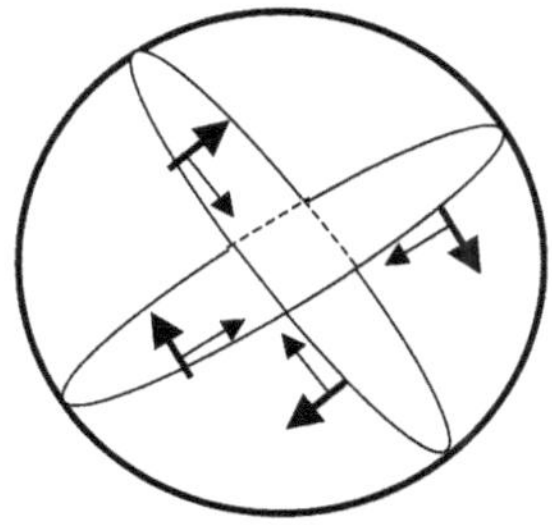

Fig. 17.2 Schematic drawing of the spin current in a spherical superconductor. The arrow perpendicular to the orbit indicates the direction of the electron's magnetic moment (opposite to its spin orientation) and the arrow parallel to the orbit indicates the direction of motion.

circulating to prevent negative electrons from 'falling' towards the interior of the superconductor due to the force exerted by the electric field on them. In the same way that the earth does not 'fall' towards the sun because it orbits around the sun. How can there be a current without an accompanying magnetic field that results from electric currents? Very simple. Besides charge, electrons have spin, in two orientations, let's call then up and down. If electrons with spin up orbit in one direction, and those with spin down orbit in opposite direction, there is no net charge current, there is a spin current, and no magnetic field is generated. Figure 17.2 shows schematically the spin current that the theory of hole superconductivity predicts exists in all superconductors, associated with the inhomogeneous charge distribution of Fig. 17.1. The orientation of the orbit relative to the spin orientation is determined by what is called the 'spin–orbit interaction', that is well known to operate in microscopic atoms. According to BCS, it plays no role in superconductors. In contrast, the theory of hole superconductivity says that a superconductor is a giant atom and the spin–orbit interaction plays an important role just as in microscopic atoms. The intrinsic magnetic moment of the electron, in opposite direction to its spin, interacts with the electric field that results from the inhomogeneous charge distribution, giving rise to the currents shown in Fig. 17.2.

In the following chapters, we will explore the consequences of this physics in more detail.

References

[1] J. E. Hirsch, Consequences of charge imbalance in superconductors within the theory of hole superconductivity, *Phys. Lett. A* **281**, 44 (2001).
[2] J. E. Hirsch, Absence of Josephson coupling between certain superconductors, *Europhys. Lett.* **109**, 67005 (2015).

Chapter 18

Holes, charge expulsion and internal electric field

As we saw in Chapter 16 (Fig. 16.13), the theory of hole superconductivity predicts that the kinetic energy of charge carriers decreases when there are more holes in the vicinity of a given hole. This is also true globally in the band: the kinetic energy per electron for all electrons in the band decreases when there are more holes in the band, or equivalently when there are fewer electrons in the band, much more so than what is predicted by traditional 'rigid band models' where the hopping amplitude is independent of density, i.e. $\Delta t = 0$. This causes the system to want to expel electrons. If it is an isolated system, the best it can do is to expel electrons to the surface, where they will pile up. Numerical calculations with the model described in Chapter 16 show this effect, [1], as Fig. 18.1 shows.

As we saw in Chapter 16, the interaction Δt increases the more negative the atom is, or in other words, the smaller the effective nuclear charge Z is. Now, in the normal metal, there cannot be charge expulsion at a macroscopic level because, as we said before, there cannot be an electric field in the interior of a metal. So, the Coulomb interaction counteracts the tendency to expel electrons, and it doesn't happen. Instead, when the metal becomes superconducting, negative charge expulsion can take place.

However, we need to take into account that at finite temperature in the superconducting state, there is coexistence of normal electrons

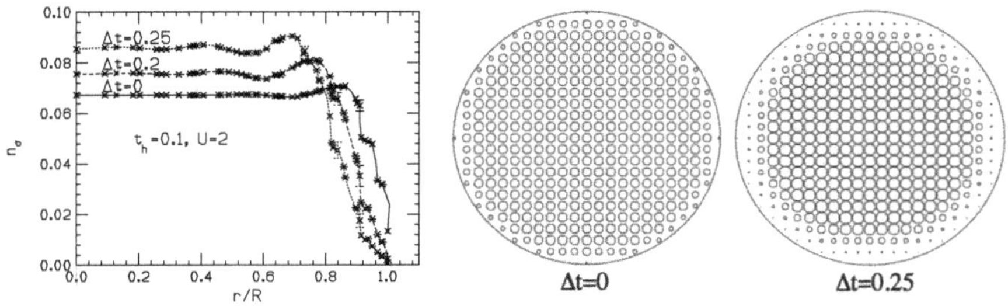

Fig. 18.1 Fraction of holes per site in a cylinder as function of the distance to the central axis of the cylinder, r, divided by the cylinder radius, R. There are more holes, i.e. fewer electrons, in the interior than near the surface. The inhomogeneity increases when the interaction Δt increases. In the figures to the right, the diameter of the circles is proportional to the hole density.

and electrons in the superfluid, i.e. electrons that conduct without resistance. The proportion of normal electrons goes to unity when the temperature approaches T_c and the superfluid density goes to zero, and goes to zero when the temperature goes to zero and all the electrons join the superfluid. At finite temperature, the normal electrons (what's called the normal fluid) screen internal electric fields as in the normal metal, so there is no macroscopic electric field in the interior. It is only at zero temperature that the total charge configuration is as shown in Fig. 17.1. At finite temperature that is the configuration of the superfluid charge, but it is compensated by an opposite configuration of the normal charge.

When the metal goes from the normal to the superconducting state and the superfluid forms, there is an expulsion of superfluid from inside out, and there is an 'impulsion' of normal fluid from outside in to partially compensate the cost in electric potential energy. The radially moving carriers acquire azimuthal velocity due to the magnetic Lorentz force, as explained in Chapter 10. These two processes, expulsion of superconducting charge and impulsion of normal charge, are essential to understand the two mysteries associated with the Meissner effect: how is the supercurrent generated, and how does the body acquire mechanical momentum in the opposite direction. The latter occurs because the normal fluid flowing inward transmits its azimuthal momentum to the ions. I understood this

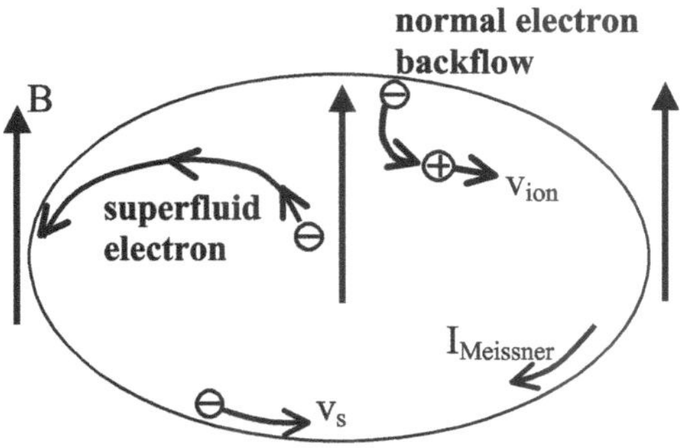

Fig. 18.2 When the metal becomes superconducting, superfluid flows from inside out and normal fluid counterflows from outside in. This explains how the Meissner current acquires its azimuthal momentum, and how the body acquires azimuthal momentum in opposite direction [2], due to the action of the Lorentz force. We will see it in detail later.

qualitatively back in 2008 [2], as Fig. 18.2 shows, but it took me several more years to understand it in detail, as we will see later.

The mathematical description of the inhomogeneous charge distribution of the superfluid is naturally done with a small modification of London's equations [3]. Recall that we said in Chapter 2 that half of London's equations are not right, specifically, Eq. (6.4). It has to be replaced by

$$m_e \frac{\partial \vec{v}_s}{\partial t} = e(\vec{E} + \vec{\nabla}\phi) \tag{18.1}$$

where ϕ is the electric potential. In the absence of external fields and in a static situation,

$$\vec{E} = -\vec{\nabla}\phi \tag{18.2}$$

and Eq. (18.1) says that there is no current in the superconductor, as it should be. (Note for the experts: Eq. (18.1) is not incompatible with Newton's equations, as it appears, because the time derivative is partial, not total). The electrostatic potential is determined by the equation

$$\phi(\vec{r}) - \phi_0(\vec{r}) = -4\pi\lambda_L^2(\rho(\vec{r}) - \rho_0) \tag{18.3}$$

where $\rho(\vec{r})$ is the charge density and $\phi_0(\vec{r})$ is the electric potential that would result from a uniform charge distribution $\rho_0 > 0$ in the interior of the superconductor. This equation, together with the condition that the net charge of the superconductor is zero, and together with the geometry of the body and the boundary conditions, determines quantitatively the charge distribution in the superconductor [3], as shown in Fig. 17.1. The excess negative charge resides in a layer of thickness λ_L adjacent to the surface, and has magnitude

$$\rho_- = \frac{R}{2\lambda_L} \tag{18.4}$$

for a cylinder of radius R, and

$$\rho_- = \frac{R}{3\lambda_L} \tag{18.5}$$

for a sphere of radius R. The maximum electric field near the surface has magnitude

$$E_m = -4\pi\lambda_L\rho_- \tag{18.6}$$

the same for both geometries.

From Eq. (18.3), we derive a relativistically covariant electrodynamics, where several physical quantities obey exactly the same differential equation [3]. For example, for the charge density

$$\nabla^2(\rho - \rho_0) = \frac{1}{\lambda_L^2}(\rho - \rho_0) + \frac{1}{c^2}\frac{\partial^2(\rho - \rho_0)}{\partial t^2} \tag{18.7}$$

and the same equation is valid replacing ρ by $\vec{B}$ (with $\vec{B}_0 = 0$), by $\vec{E}$, with $\vec{E}_0$ the electric field resulting from a uniform charge distribution ρ_0, by $\vec{J}$ (current density) with $\vec{J}_0 = 0$, etc. The fact that the equations we obtain are so symmetric, compact, and aesthetically appealing suggests that they describe physical reality.

It is interesting to note that these equations, that are not what is commonly known as London's equations, were originally postulated, for the particular case $\rho_0 = 0$, by the London brothers themselves in 1935, as one possible description of the electrodynamics of superconductors. But they discarded them a year later because of an experiment that H. London did in 1936 [4] that seemed to indicate that they were not valid, and because it appeared unphysical to have the possibility of electric fields inside superconductors [5]. The reality is that the H. London's experiment of 1936 couldn't really tell anything about this because it didn't reach sufficiently low temperatures. The reasons for why in fact it is physical to expect that a superconductor at very low temperatures may have an electric field in the interior were discussed in Chapter 17.

Consequences of this electrodynamic theory that are different from the conventional theory are, for example: (1) At sufficiently low temperatures, applied static electric fields should penetrate the superconductor a distance λ_L, that is several hundred Å, instead of a fraction of Å as the conventional theory predicts. (2) For bodies without spherical or cylindrical symmetry, the inhomogeneous charge distribution can give rise to electric fields in the exterior of the body, if the dimensions of the body are sufficiently small ($\sim 1\mu m$) and the temperature sufficiently low. These consequences are experimentally measurable with existing techniques, but haven't been measured yet. Examples of the electric field distribution expected in the vicinity of small superconducting particles at low temperatures are shown in Fig. 18.3 [6, 7].

It is interesting that in the theory of hole superconductivity, there is a parallel between the microscopic physics and the macroscopic

Fig. 18.3 Lines of electric field in the vicinity of superconductors of various shapes. The lines go out of regions of low surface curvature and go in in regions of high surface curvature.

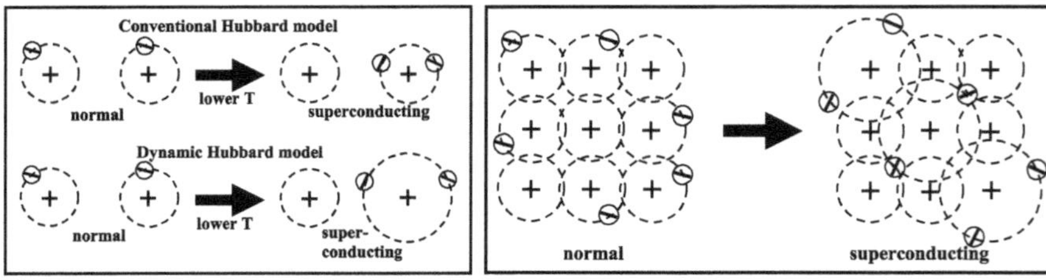

Fig. 18.4 Left panel: comparison between the conventional Hubbard model and the "dynamic Hubbard model" [8] used in the theory of hole superconductivity. Only in the dynamic model when there is pairing, the orbital expands and negative charge is expelled outward, as the right panel shows.

electrodynamics that we have been discussing, for the following reason: in the microscopic theory discussed in Chapter 16, the electron–hole asymmetry originates in the expansion of the atomic orbital when it is doubly occupied (Fig. 16.7). According to BCS theory, superconductivity occurs when electrons pair, with which I completely agree. In an extreme case of strong interactions, the pairing will occur on the same site. As the left panel of Fig. 18.4 shows, in the case when there is pairing, the orbital expands. Hence, as the right panel of Fig. 18.4 shows, there will be more expanded orbitals in the superconducting state than in the normal state, and that leads to negative charge moving out, it is expelled. In the conventional Hubbard model, as is shown in the upper left panel of Fig. 18.4, there is no orbital expansion when electrons pair and as a consequence there is no expulsion of negative charge.

In Fig. 18.5, we show again how the expansion of the orbitals, when the metal is cooled (left panel) and pairing occurs, leads to the local expansion of atomic orbitals (central panel) and to the macroscopic charge redistribution (right panel) as was also shown in Fig. 17.1. It is also notable that the expansion of the orbital leads to the existence of the Δt interaction (Chapter 16), and it in turn leads to the expulsion of negative charge as the numerical results of Fig. 18.1 show. All this is also related to lowering of kinetic energy, as we already mentioned and will analyze in more detail later. The fact that the physics from various different viewpoints converges to the same conclusion suggests that this physics describes natural

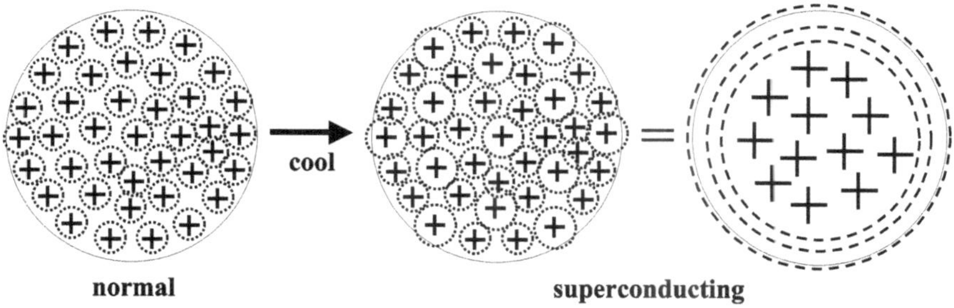

Fig. 18.5 Expansion of atomic orbitals when a normal metal is cooled (left panel) and pairing occurs (center panel) leads to the global expulsion of negative charge shown in the right panel.

reality and not the imagination of physicists as it happens with other theories.

It remains to be determined what is the amount of charge that is expelled, that in the macroscopic electrodynamics, Eqs. (18.3)–(18.7), is determined by the magnitude of the parameter ρ_0, or ρ_-. This we could only determine several years after we proposed this non-conventional electrodynamics, when we analyzed mesoscopic orbits and spin currents, as we will explain in the following chapter.

References

[1] J. E. Hirsch, Charge expulsion, charge inhomogeneity, and phase separation in dynamic Hubbard models, *Phys. Rev. B* **87**, 184506 (2013).
[2] J. E. Hirsch, The missing angular momentum of superconductors, *J. Phys. Cond. Matt.* **20**, 235233 (2008).
[3] J. E. Hirsch, Electrodynamics of superconductors, *Phys. Rev. B* **69**, 214515 (2004).
[4] H. London, An experimental examination of the electrostatic behaviour of supraconductors, *Proc. Roy. Soc. London A* **155**, 102 (1936).
[5] M. Von Laue, F. London, and H. London, Zur Theorie der Supraleitung, *Z. Phys.* **96**, 359 (1935).
[6] J. E. Hirsch, Predicted electric field near small superconducting ellipsoids, *Phys. Rev. Lett.* **92**, 016402 (2004).
[7] J. E. Hirsch, Correcting 100 years of misunderstanding: electric fields in superconductors, hole superconductivity, and the Meissner effect, *J. Supercond. Nov. Magn.* **25**, 1357 (2012).
[8] J. E. Hirsch, Dynamic Hubbard model, *Phys. Rev. Lett.* **87**, 206402 (2001).

Chapter 19

Mesoscopic orbits and spin currents

Electrons in superconductors reside in mesoscopic orbits of radius $2\lambda_L$. I understood this for the first time in 2007 [1], the 50th anniversary of the development of the BCS theory. It seems incredible that neither BCS nor anybody that works with the conventional theory knows this. Soon thereafter, I realized that John Slater already knew this in 1937, as we saw in Chapter 11.

We already discussed one argument leading to this conclusion in Chapter 11, Eqs. (11.4) and (11.5). An even simpler argument is to consider the angular momentum of the Meissner current in a cylinder of radius R and height h. The angular momentum of an electron in an orbit of radius r is $m_e v_s r$, where m_e is the mass and v_s is the velocity. The angular momentum of all the electrons in the supercurrent is

$$L_e = n_s[2\pi R\lambda_L h](m_e v_s R) \tag{19.1}$$

where n_s is the density of electrons, number of electrons per unit volume. The term in the square bracket is the volume where the supercurrent circulates: a ring of thickness λ_L adjacent to the surface of radius R has volume: perimeter $= 2\pi R$ multiplied by thickness λ_L multiplied by height h. The term in round brackets is the angular momentum of each electron in that ring, which is essentially at distance R from the cylinder axis (we assume $\lambda_L \ll R$), that is, it circulates around the cylinder in an orbit of radius R.

We can rewrite Eq. (19.1) as

$$L_e = n_s[2\pi R\lambda_L h](m_e v_s R) = n_s[\pi R^2 h](m_e v_s(2\lambda_L)). \tag{19.2}$$

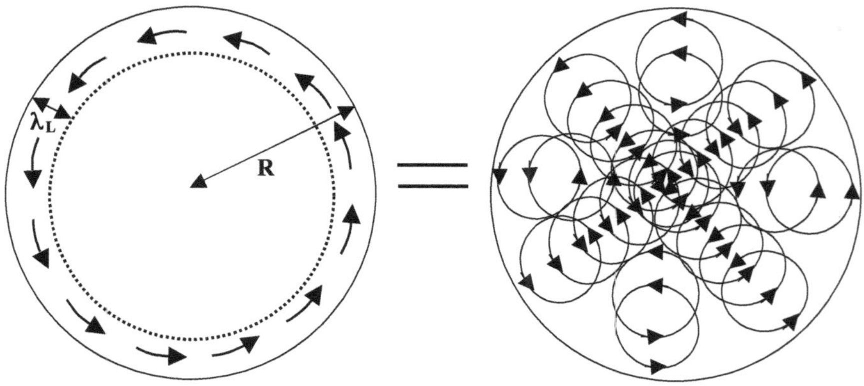

Fig. 19.1 Equivalence between the angular momentum of a current circulating in a shell of thickness λ_L next to the surface (left panel), and all the electrons circulating in orbits of radius $2\lambda_L$ (right panel).

On the right, the term in square brackets is the volume of the cylinder, the term in round brackets is the angular momentum of an electron in an orbit of radius $(2\lambda_L)$. The reader can easily verify the mathematical equivalence of both expressions in Eq. (19.2). The expression on the right tells us that the electrons throughout the volume circulate in orbits of radius $2\lambda_L$. In the interior, the velocities cancel each other out, near the surface they don't, which explains the mathematical identity. Figure 19.1 shows it schematically.

As we saw in Chapter 11, the diamagnetic susceptibility of a solid is given by Larmor's formula:

$$\chi = -\frac{n_s e^2}{4m_e c^2}\langle r^2\rangle \tag{19.3}$$

with r the radius of the electronic orbits. For a superconductor, with $r = 2\lambda_L$, we obtained (Eq. (11.5)) $\chi = -1/(4\pi)$, perfect diamagnetism. For a normal metal, the diamagnetic susceptibility is given by Landau's formula:

$$\chi_{\text{Landau}} = -\frac{1}{3}\mu_B^2 g(\epsilon_F), \tag{19.4}$$

where $\mu_B = e\hbar/2m_e c$ is the Bohr magneton and $g(\epsilon_F)$ is the density of electronic states at the Fermi level, given by $g(\epsilon_F) = 3n_s/2\epsilon_F$,

$\epsilon_F = \hbar^2 k_F^2/2m_e$, where k_F is the 'Fermi momentum' and ϵ_F is the 'Fermi energy'. That is well-known solid state physics. What nobody seems to have noticed though is that we can write Eq. (19.4) as

$$\chi = -\frac{n_s e^2}{4m_e c^2}(k_F^{-2}). \tag{19.5}$$

Comparing with Eq. (19.3), this implies that in the normal metal, electrons move in microscopic orbits of radius k_F^{-1}. The relation between k_F and the density of electrons is $n_s = k_F^3/3\pi^2$. In a normal metal, $k_F^{-1} \sim 1$ Å, which gives a Landau susceptibility $\chi_{\text{Landau}} \sim -10^{-6}$, as observed experimentally.

This implies that we can understand the transition from metal to superconductor, where the diamagnetic susceptibility increases by a factor ~100,000, from $\chi_{\text{Landau}} \sim -10^{-6}$ to $\chi = -1/4\pi$, as an *expansion* of electronic orbits from radius $k_F^{-1} \sim 1$ Å to radius $2\lambda_L \sim 800$ Å!

What is important about this interpretation is that it provides a dynamical explanation of how the electrons acquire the speed of the Meissner current, which, as we said earlier in (Eq. (6.13)), is

$$v_s = -\frac{e\lambda_L}{m_e c}B \tag{19.6}$$

in the presence of magnetic field B. As the orbit expands, there is radially outward motion of charge, and the Lorentz force imparts its azimuthal velocity, as we discussed in Chapter 10. We show this in Fig. 19.2. Figure 19.2 also shows the mathematical steps to obtain this result, using simply Newton's equations and rules of vector algebra.

We conclude from this reasoning that in the transition from the normal to the superconducting state, electronic orbits expand from radius k_F^{-1} to radius $2\lambda_L$, and in the presence of a magnetic field the Lorentz force imparts precisely the azimuthal velocity needed so that the resulting current cancels the applied magnetic field, resulting in zero magnetic field in the interior of the superconductor. Certainly, BCS theory doesn't say anything like that. John Slater realized that in the superconductor, there are mesoscopic orbits of

$$\vec{F} = \frac{e}{c}\vec{v} \times \vec{B} + \vec{F}_r = m_e \frac{d\vec{v}}{dt}$$

$$\frac{d}{dt}(\vec{r} \times \vec{v}) = -\frac{e}{2m_e c}(\vec{r} \cdot \vec{v})\vec{B} = -\frac{e}{2m_e c}\frac{d}{dt}(r^2)\vec{B}$$

$$=> v_\phi = -\frac{er}{2m_e c}B \qquad r=2\lambda_L \qquad ==> v_\phi = \frac{e\lambda_L}{m_e c}B$$

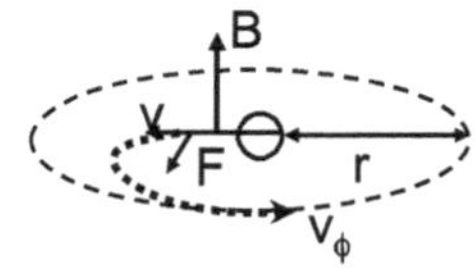

Fig. 19.2 Expansion of electronic orbit from a microscopic radius (assumed 0) to radius $r = 2\lambda_L$. The Lorentz force imparts azimuthal velocity v_ϕ equal to the velocity of electrons in the Meissner current Eq. (19.6).

radius $\sim 2\lambda_L$, but did not realize that the expansion of the microscopic orbit (of radius k_F^{-1}) to mesoscopic radius explains dynamically how the Meissner current is generated. This paper by Slater of 1937 (Ref. 8 in Chapter 11) has only 27 citations, of which 13 are mine and 2 are from Bardeen in works he did before BCS. That is not much for Slater who has *H*-index of 55 and 24,563 total citations, on average almost 200 citations for each of the 127 articles he published, very impressive for a scientist who started publishing in 1924.

Furthermore, it is natural to conclude that this expansion of orbits occurs both in the presence and in the absence of applied magnetic field. The magnetic field does not produce the orbit expansion, it simply acts in the process of expansion to generate the Meissner current, as we saw above.

The fact that electrons in the superconducting state reside in orbits of radius $2\lambda_L$ is totally alien to BCS. Note that the expression for λ_L contains the speed of light, c. In BCS theory, the speed of light doesn't enter, unless a magnetic field is applied, in that case the response of the system depends on c. But in the absence of the magnetic field, no property of the superconductor depends on c according to BCS, for that reason it is said that the theory is "non-relativistic". The fact that in reality the radius of the orbits in the absence of magnetic field depends on c informs us that relativity plays an essential role in superconductivity, contrary to what BCS tells us. This is because the spin-orbit interaction, a relativistic effect, plays an essential role, as we will see in what follows.

What happens when orbits expand if there is no applied magnetic field?

Here, a new and very interesting effect occurs that we have called 'spin Meissner effect' [1]. It goes like this: the intrinsic magnetic moment of the electron interacts with the electric field generated by the positive ionic charge that neutralizes the negative charge of the superfluid, and that interaction also gives rise to an azimuthal force when the orbit expands. The electric field generated by the ionic charge is $\vec{E} = 2\pi\rho\vec{r}$, where $\rho = |e|n_s$ is the ionic charge density that neutralizes the negative charge of the superfluid. The result is [1] that the electron 'feels' an effective magnetic field, given by

$$\vec{B}_\sigma = 2\pi n_s \vec{\mu} \tag{19.7}$$

with

$$\vec{\mu} = \mu_B \hat{z} = \frac{e\hbar}{2m_e c}\hat{z}, \tag{19.8}$$

the magnetic moment of the electron. $\bar{z}$ is the direction of the cylinder axis and $\sigma = \pm 1$ is the direction of the electron spin. The same calculation as in Fig. 19.2 with B_σ instead of B yields the result (using Eq. (6.10b)) that when the orbit expands to radius $2\lambda_L$ the electron acquires azimuthal velocity

$$\vec{v}_\sigma^0 = -\frac{\hbar}{4m_e\lambda_L}\vec{\sigma} \times \hat{n}, \tag{19.9}$$

where $\vec{\sigma} = \sigma\hat{z}$, $\sigma = \pm 1$ and $\hat{n}$ is the outward normal to the cylinder surface. Quantitatively, for example, for a typical value $\lambda_L = 400\,\text{Å}$, $v_\sigma^0 = 72,395\,\text{cm/s}$.

This implies that when the expansion of orbits occurs in the absence of the magnetic field, half of the electrons, with spin pointing down, end up orbiting in counterclockwise direction (seen from the top), and the other half, with spin pointing up, end up orbiting in clockwise direction. In the interior, these currents cancel, and within a layer of thickness λ_L adjacent to the surface, a 'spin current' circulates, where half of the electrons go one way, the other half go the opposite way, giving zero net charge current. We show this in Fig. 19.3.

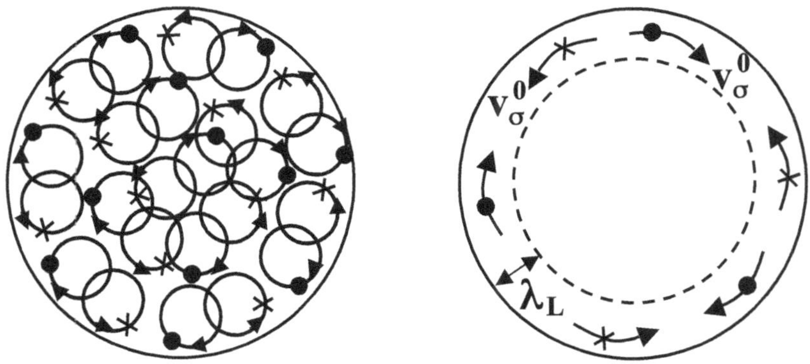

Fig. 19.3 Ground state of a superconducting cylinder, seen from the top. Electrons move in mesoscopic orbits of radius $2\lambda_L$. The crosses indicate spin pointing into the paper, the black circles indicate spin pointing out of the paper. The right panel shows the superposition of these orbits, giving rise to a spin current within a layer of thickness λ_L adjacent to the surface. The orbital angular momentum of the electron is $\hbar/2$, pointing in opposite direction to the electron spin direction (see text).

Now, we need to return to the analysis of the orbit expansion in the presence of an external magnetic field B, and we simply use superposition: the Lorentz force that acts when the orbit expands is due to the sum $\vec{B} + \vec{B}_\sigma$. If the field $\vec{B}$ points in direction $+\hat{z}$, it accelerates the electron of spin down in counterclockwise direction and decelerates the spin up electron in clockwise direction. The net velocity acquired is

$$\vec{v}_\sigma = -\frac{\hbar}{4m_e\lambda_L}\vec{\sigma} \times \hat{n} - \frac{e\lambda_L}{m_e c}\vec{B} \times \hat{n}. \tag{19.10}$$

The electrons that are slowed down stop when the applied magnetic field reaches a critical value B_c that is simply obtained from Eq. (19.10) putting $\vec{v}_\sigma = 0$, and is given by

$$B_c = -\frac{\hbar c}{4e\lambda_L^2} \tag{19.11}$$

that can also be written as

$$B_c = \frac{\phi_0}{\pi(2\lambda_L)^2}. \tag{19.12}$$

$\phi_0 = hc/(2e)$ is what is called the magnetic flux quantum. The magnetic flux is the product of the magnetic field times the area it goes through, and it is quantified in units of ϕ_0. Equation (19.12) says that when the magnetic field is B_c, there is one unit of flux quantum through an orbit of radius $2\lambda_L$.

It is remarkable that the expression (19.12) is the same as what BCS gives for the critical magnetic field [2] for which the superconducting state that completely excludes the magnetic field is destroyed, and magnetic vortices can enter the superconductor. So, we conclude that when the spin current stops ($v_\sigma = 0$) for one of the spin components, the superconducting state is destroyed. That is, the superconducting state *requires* that a spin current exists. Figure 19.4 shows this current schematically in the absence and in the presence of an applied magnetic field B. The critical velocity of the charge current is $v_s = v_\sigma^0$, for example, ~72,000 cm/s for $\lambda_L = 400$ Å.

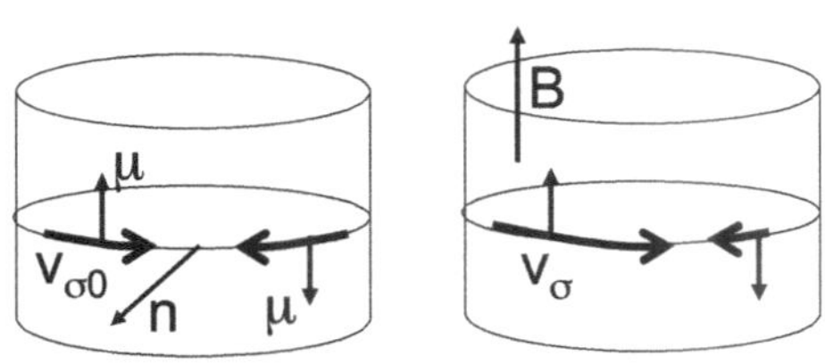

Fig. 19.4 Spin current close to the surface of a cylindrical superconductor. The magnetic moment μ points in the direction opposite to the electron spin. In the presence of an applied magnetic field (right panel), one of the components speeds up and the other slows down.

But we haven't yet said what is the most remarkable consequence of this. If we calculate the angular momentum of electrons in these orbits of radius $2\lambda_L$ that give rise to the spin current, we obtain

$$\ell = m_e v_\sigma^0 (2\lambda_L) = m_e \frac{\hbar}{4 m_e \lambda_L}(2\lambda_L) = \text{........................} = \frac{\hbar}{2} \text{ !!!} \tag{19.13}$$

This result is, in one word, *spectacular*. Let's see why.

According to quantum mechanics, the angular momentum of electrons in microscopic atomic orbits is quantized in multiples of $\hbar$, Planck's constant. In the semiclassical calculation that we did to obtain the velocity v_σ^0, Eq. (19.9), we didn't use quantum mechanics. Planck's constant $\hbar$ appeared because it gives the magnitude of the intrinsic magnetic moment of the electron, Eq. (19.8). But the calculation only used Newton's equations and classical electrodynamics, and there was no reason to expect that the angular momentum of electrons in the mesoscopic orbits of radius $2\lambda_L$, that we inferred have to exist to explain the Meissner effect, acquired by the interaction of the magnetic moment of the electron with the ions when the orbit expands, would be quantified in a simple integer multiple of $\hbar$, and much less in the simple fraction $\hbar/2$. To obtain this result, the assumption of charge neutrality, used to obtain Eq. (19.7), was crucial.

Recall that the intrinsic angular momentum of the electron, its spin, is $\hbar/2$. Even though it is sometimes visualized as an intrinsic rotation of the electron around its own axis, quantum mechanics textbooks tell us that it is not real, that an electron is a point particle and cannot rotate around an axis centered on itself, that the spin is an abstract concept. Instead, here, we find that the angular momentum of these mesoscopic orbits in the superconducting state *reflects* the intrinsic angular momentum of the electron. It is as if the electron 'expanded' from being a point particle to being a ring of radius $2\lambda_L$!

We will see more about this 'expansion' of the electron in Chapter 21. I don't really know what it means, but cannot avoid thinking that this totally unexpected result [1] reflects a deep principle of nature. It cannot be a coincidence.

Figure 19.3 shows the ground state of a cylindrical superconductor as seen from the top. Electrons in the superfluid move in these mesoscopic orbits in the absence of applied external fields. It is truly, like London had guessed (see Chapter 11), "*a kind of kinetic equilibrium for the so-called zero point motion, which may roughly be characterized as defined by the minimum average total (potential + kinetic) energy*". Indeed, the resulting spin current is a *macroscopic* 'zero-point motion', as one would naturally expect: microscopic quantum systems always have microscopic zero-point motion, right? Then it is natural to expect that a macroscopic quantum system, that everybody agrees a superconductor is, should have macroscopic zero-point

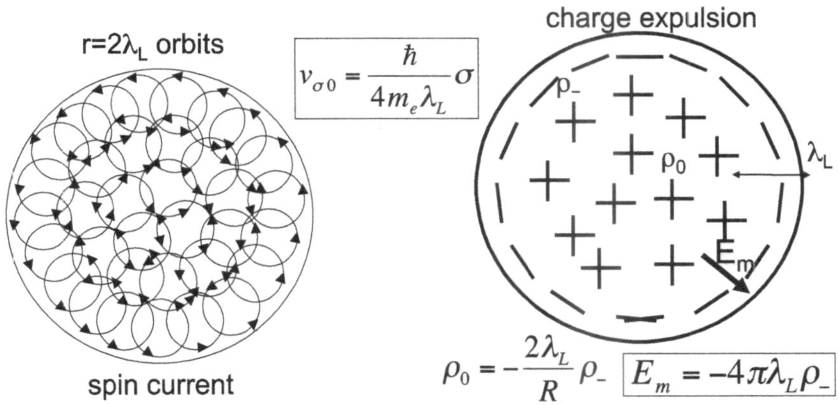

Fig. 19.5 Two new properties of superconductors, not predicted by BCS. The left panel shows the mesoscopic orbits giving rise to the spin current, the right panel shows the inhomogeneous charge distribution.

motion. Why didn't anybody think about this? Because it is not predicted by BCS, in which everybody believes.

So, we now have two new properties of superconductors that are certainly not described by BCS theory: the existence of mesoscopic orbits giving rise to spin current, and the macroscopic charge inhomogeneity that we discussed in Chapter 18. We show them juxtaposed in Fig. 19.5. It is intuitively clear that these two properties are related: expansion of the orbit leads to charge moving out. But what is the quantitative relationship?

As will be remembered from Chapter 18, the electrodynamic equations that give rise to charge inhomogeneity depend on a parameter ρ_0 that gives the density of positive charge in the interior, or equivalently on a parameter ρ_- that gives the density of negative charge near the surface. For the cylinder, the relation between these two quantities is given in Fig. 19.5, as is the relation between ρ_- and the maximum electric field in the interior, E_m. The electric field at distance r from the axis of the cylinder is given simply by

$$E = \frac{r}{R}E_m. \tag{19.14}$$

The reasoning we used was the following [3]. Extending the electrodynamic relations to four dimensions, which means incorporating

Einstein's theory of relativity, we found that the application of a magnetic field changes the charge density for each spin distribution: the one that is accelerated increases its charge density, the one that is decelerated decreases it. This allowed us to deduce a relation between the charge density near the surface and the velocity of the spin current, given by Eq. (19.9), and we found the incredibly simple relationship [3]:

$$\rho_- = en_s \frac{v_\sigma^0}{c}. \tag{19.15}$$

In other words, the excess negative charge is given by the charge density of the superfluid (en_s) multiplied by the ratio between the speed of the spin current v_σ^0, Eq. (19.9), and the speed of light, c. Recall that we said in an example $v_\sigma^0 \sim 70,000\,\text{cm/s}$, since $c = 3 \times 10^{10}\,\text{cm/s}$ we have $v_\sigma^0/c \sim 1/500,000$. That is, the excess negative charge is a small fraction of the total charge of the superfluid en_s.

The resulting maximum electric field is $E_m = -4\pi\lambda_L\rho_-$, determined by electrostatics so that the electric field outside the cylinder is 0, as it should be if there is charge neutrality. This yields, using Eqs. (19.9) and (6.10b),

$$E_m = -4\pi\lambda_L\rho_- = -4\pi\lambda_L en_s \frac{\hbar}{4m_e\lambda_L c} = -\frac{\hbar c}{4e\lambda_L^2}. \tag{19.16}$$

Notably and unexpectedly, we obtain the same expression for the maximum internal electric field E_m as for the critical magnetic field that stops one of the components of the spin current, Eq. (19.11).

In the system of units that we are using (cgs), electric and magnetic fields have the same units, that's why Eq. (19.16) can have exactly the same form as Eq. (19.11). If instead of expressing the electric field in that system of units, we express it in the SI system more used in engineering, the electric field is measured in Volts/cm (V/cm). The relation with the magnetic field, which is measured in Gauss (G), is 1 G = 300 V/cm. For the example we were considering, with $\lambda_L = 400\,\text{Å}$, we obtain $E_m = 308,300\,\text{V/cm}$, $B_c = 1028\,\text{G}$.

There are many other very remarkable properties that follow from these relations that we will not elaborate on here. The interested reader can find more information on the subject in the paper [3].

Concluding this chapter, let us remember that in the same paragraph of Fritz London that we mentioned above and in Chapter 11, he also said *"If the resultant forces are sufficiently weak and act between sufficiently light particles, then the structure possessing the smallest total energy would be characterized by a good economy of the kinetic energy"*. As almost always, he was right. We already discussed a bit about kinetic energy in Chapter 16, and will focus on it in the following chapter.

References

[1] J. E. Hirsch, Spin Meissner effect in superconductors and the origin of the Meissner effect, *Europhys. Lett.* **81**, 67003 (2008).

[2] M. Tinkham, *Introduction to Superconductivity*, McGraw Hill, New York, 1996.

[3] J. E. Hirsch, Electrodynamics of spin currents in superconductors, *Ann. Phys.* (Berlin) **17**, 380 (2008).

Chapter 20

Kinetic or potential energy? Liquid Helium provides the answer

In a paper I wrote in 2010 [1], I said that understanding how the Meissner effect works is similar to understanding how a battery works. Figure 20.1 explains why.

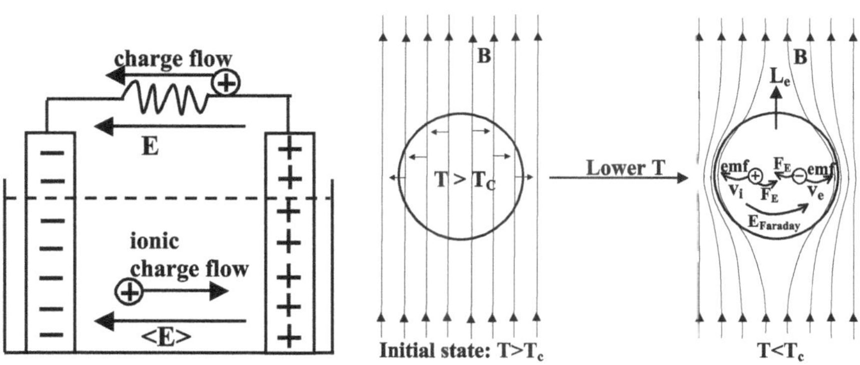

Fig. 20.1 Analogy between voltaic cells and superconductors. In a voltaic cell (left panel), there is a flux of positive ions in the interior in the direction of the positive electrode, apparently against a repulsive electric force. In a superconductor that expels magnetic field (right panel), electrons and ions move in opposite direction to the electric force F_E acting on them by the Faraday field.

On the left panel, there is a voltaic cell (battery). The electrode on the right is positive and the one on the left is negative. If we connect them with a wire, positive charge goes from right to left, in reality what happens is that negative electrons go from left to right. There is however no charge accumulation at the electrodes. The reason is because inside the battery there is a flow of positive ions from left to right that compensates the charge flowing in the wire.

Now inside the cell there is an electric field that on average points in the same direction as outside, that is from right to left, called $\langle E \rangle$ in the figure. This is because the integrals of the electric field along a path outside the cell or inside connecting the two electrodes have to be equal, because $\vec{E}$ is what is called a 'conservative field'. Therefore, the ions that flow inside the cell move in direction opposite to the one the electric field is pushing them to go. How is that possible?

Exactly the same situation occurs in the Meissner effect in a superconductor, as we saw earlier and is illustrated on the right panel of Fig. 20.1. As the magnetic field is expelled, a Faraday electric field is generated that pushes electrons and ions to move in opposite direction to the one they actually move.

The question is then the same in both cases. How do electric charges manage to move in opposite direction to the one dictated by the electric field? There has to be another force, let's call it 'emf' ('electromotive force'). In that article [1], I proposed that the emf is the same for voltaic cells and for superconductors.

Fortunately, for voltaic cells it is known where the emf comes from, even though it is not very well explained in the books. We show it in Fig. 20.2.

The left panel shows the electric field in the interior of the cell. It is highly inhomogeneous. Over most of the region between the electrodes it points to the right (E_2), and there it propels the ions to move from left to right. But there are small regions on both sides where it is much bigger and points from right to left, in regions called 'double layers', of microscopic thicknesses d_1 and d_3.

Where do these 'double layers', with internal electric field E_1 and E_3, come from? From the separation of electric charge in those regions, like the right panel of Fig. 20.2 shows. Chemistry books explain this charge separation with a variety of terms: "oxidation and reduction potentials", "contact potentials", "free energy",

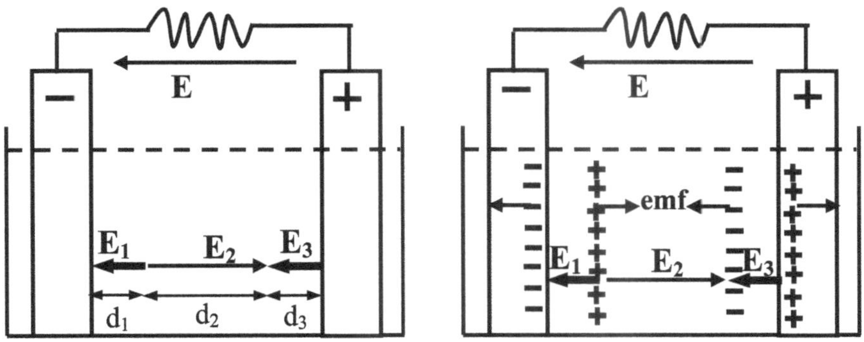

Fig. 20.2 Explanation of how a voltaic cell works. Even though on average the electric field goes from right to left, over most of the space between the electrodes it goes from left to right (E_2), moving the ions as Fig. 20.1 left panel shows. Only in the 'double layers' of microscopic thickness d_1 and d_3 does the electric field point in opposite direction (and is enormous). The 'emf' that creates these double layers is quantum kinetic energy.

"electrochemical potential", etc. I say it is, simply, *quantum kinetic energy.*

To create the double layer clearly costs electrostatic energy given by (for the left double layer)

$$U_E = \int \frac{E^2}{8\pi} d^3r \sim \frac{E_1^2}{8\pi} d_1 A_1 \tag{20.1}$$

where A_1 is the area of the electrode. Obviously there is an electrostatic force that pushes in the direction of reducing d_1 and U_E, and there is an 'emf' that pulls in the opposite direction. In a quantum system, the total energy is given by

$$E = \langle\Psi|K|\Psi\rangle + \langle\Psi|U|\Psi\rangle \tag{20.2}$$

where $|\Psi\rangle$ is the wavefunction, and K and U are operators of kinetic and potential energy. Clearly the electrostatic energy Eq. (20.1) comes from the second term in Eq. (20.2). Therefore, if the double layer forms spontaneously, lowering the total energy E, it must be because the first term in Eq. (20.2), the kinetic energy, decreases more than what the second term increases.

Generically, the quantum kinetic energy of a particle of mass m is given by the expression (cf Eq. (17.1))

$$\epsilon_{\text{kin}} \sim \frac{\hbar^2}{2m\lambda^2} \tag{20.3}$$

where λ is a measure of the region within which the particle is confined. When λ expands, the quantum kinetic energy decreases. For example, in a voltaic cell where the negative electrode is Zn, it is diluted in a solution of SO_4Zn as an ion Zn^{++} and it leaves two electrons at the electrode, creating a double layer with ions Zn^{++} close to the negative electrode. The reason for why Zn becomes ionized is because the electrons expand their wavefunction in going from the atom Zn to the metallic electrode, that way lowering their quantum kinetic energy. That is the origin of the emf in the battery.

Analogously, the reason for why in the superconductor charges move in opposite direction than what the Faraday field says, and negative charge is expelled from the interior outward against electrostatic forces that want to keep it inside, is reduction of quantum kinetic energy due to expansion of the electronic wavefunction when the system enters the superconducting state. As we discussed in Chapter 19, electronic orbits expand from microscopic radius k_F^{-1} to radius $2\lambda_L$ when the normal metal becomes superconducting. According to Eq. (20.3) and as Fig. 20.3 shows, this results in a lowering of quantum kinetic energy.

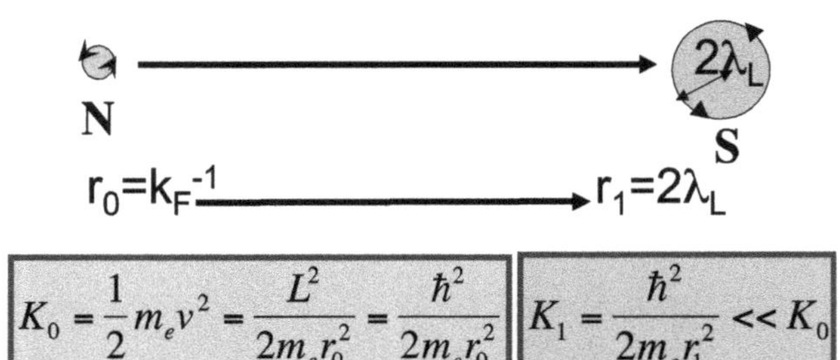

Fig. 20.3 In the transition from the normal to the superconducting state, electronic orbits expand and as a consequence the quantum kinetic energy K decreases.

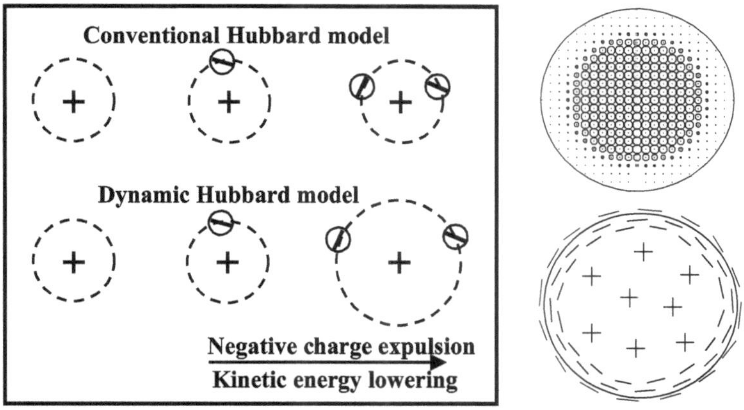

Fig. 20.4 Left panel: the atomic orbital expands when it is occupied by two electrons, and this causes a decrease of the quantum kinetic energy. Right panel: the upper circle shows the hole occupation in the interior, that gets smaller toward the surface, that is, there are excess electrons near the surface; the lower circle shows the resulting charge distribution.

As was mentioned earlier, also in the microscopic model, there is an expansion of the atomic orbital when two electrons occupy the same orbital, as Fig. 20.4 shows. This brings about a lowering of kinetic energy. The inhomogeneous charge distribution costs potential energy, and this is compensated by the lowering of quantum kinetic energy. Instead, in the conventional Hubbard model, the quantum kinetic energy does not change when two electrons occupy the same orbital, and there isn't inhomogeneous charge distribution.

The reduction of quantum kinetic energy can be measured in optical experiments. In 1992, I did a calculation predicting this [2, 3], and 10 years later it was detected experimentally in the cuprates [4–6]. In Fig. 20.5, the lower right panel shows how the optical absorption changes when the metal becomes superconducting. There is a lowering at high frequency, in the visible spectrum, which causes the color of the metal to change slightly, and an increase at low frequency. In the microscopic theory of Chapter 16, this is a direct consequence of the term Δt in the model.

Instead, within BCS theory, this is qualitatively different: the quantum kinetic energy increases when the system becomes a superconductor, and it is instead the potential energy that decreases.

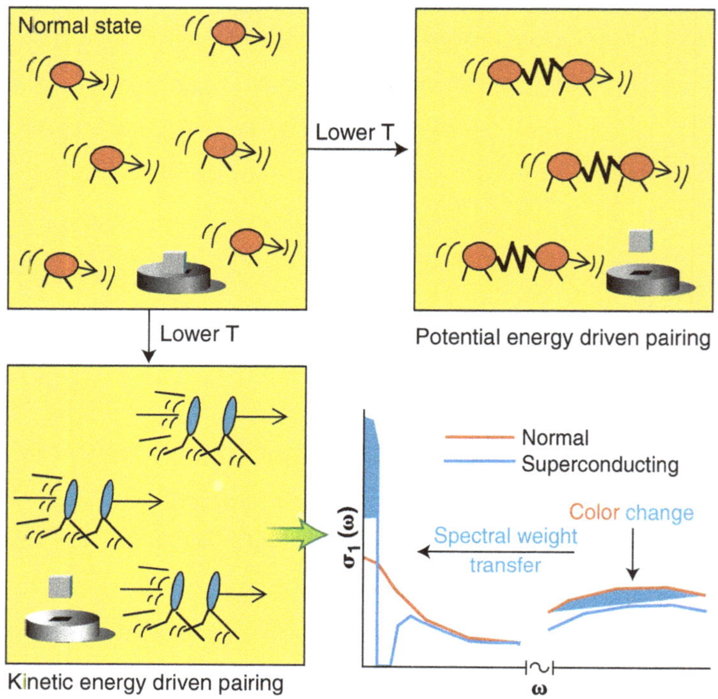

Fig. 20.5 The lowering of quantum kinetic energy is measured in optical experiments [4–6]. The figure contrasts the behavior of superconductors within BCS theory, where the charge carriers pair without changing their mobility (upper right panel), versus the behavior within the theory of hole superconductivity, where carriers propagate more easily and reduce their quantum kinetic energy when they pair (lower left panel). Only in the latter case the changes in optical absorption $\sigma_1(\omega)$ shown on the lower right panel are predicted.

Furthermore, the theory does not predict any 'change in color' when a metal becomes superconducting.

It is interesting to compare this physics of superconductors with what occurs in another macroscopic quantum system: liquid helium, ^{4}He. ^{4}He undergoes a phase transition at temperature $T_\lambda = 2.17\,\mathrm{K}$, the famous 'lambda transition'. The name derives from the fact that the specific heat as function of temperature has a behavior that resembles the greek letter λ (inverted). At temperatures lower than T_λ part of the liquid becomes superfluid, that means it flows without any viscosity. In this system it is known, for many reasons both experimental and theoretical, that the transition is associated with lowering of quantum kinetic energy, the same we say occurs in superconductors, contrary to what BCS says. Figure 20.6 shows the behavior of kinetic and potential energy in superconductors as a function of temperature for the theory of hole superconductivity (left panel) and for conventional BCS theory (right panel) [7].

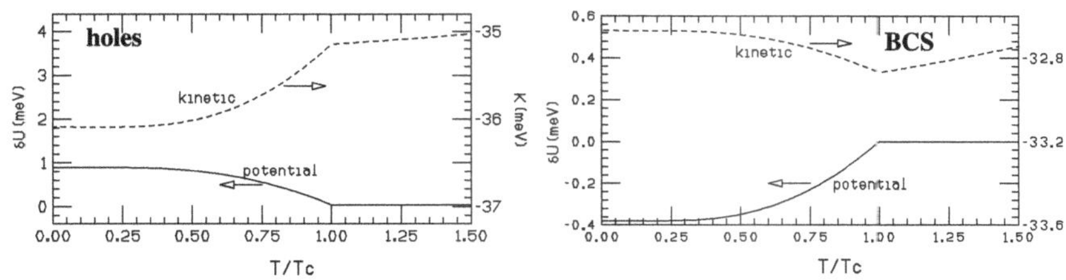

Fig. 20.6 Kinetic and potential energy of a superconductor as function of temperature for a system described by the theory of hole superconductivity (left panel) and for one described by BCS theory (right panel).

And Fig. 20.7 shows the results of numerical calculations with quantum models of ^{4}He [8]. The kinetic and potential energies as function of temperature show analogous behavior to the one given by the theory of hole superconductivity (left panel of Fig. 20.6) and qualitatively different from what BCS theory predicts for superconductors (right panel of Fig. 20.6).

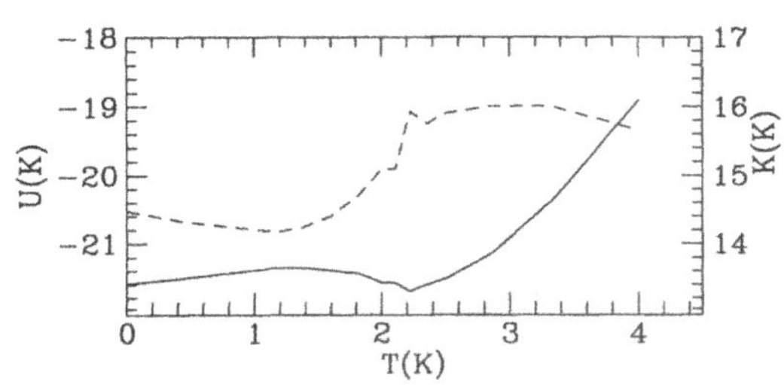

Fig. 20.7 Kinetic energy (dashed line) and potential energy (full line) as function of temperature for a model of liquid helium [8]. Note the similarity with the left panel of Fig. 20.6, and the qualitative difference with the right panel.

But Fig. 20.7 is a theoretical calculation. Let us look at experimental evidence that the transition to the superfluid state in ^{4}He is associated with lowering of quantum kinetic energy, in Fig. 20.8.

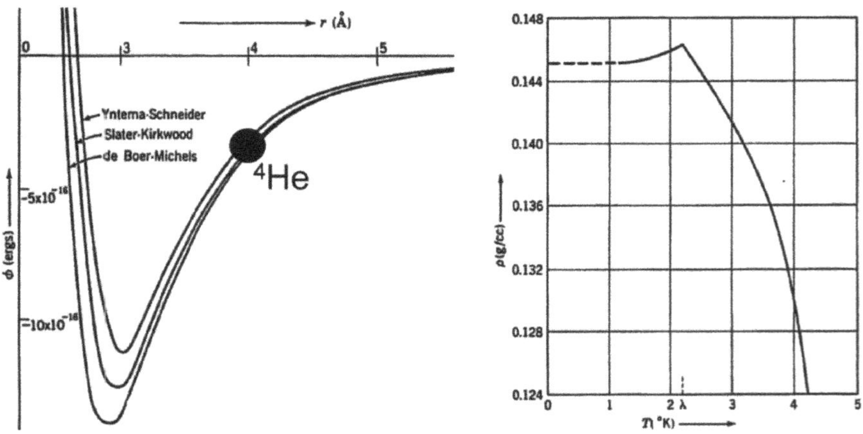

Fig. 20.8 Left panel: interaction potential energy between two He atoms as function of their distance. The distance corresponding to the density of liquid ^{4}He near the λ point is indicated by the black circle. Right panel: density of ^{4}He as function of temperature.

The left panel shows the interaction potential energy between two He atoms as function of the distance between the atoms, measured by various authors in experiments with gases. They all coincide that the minimum in potential energy occurs when the distance between the atoms is approximately 3 Å (center to center). Now the density of liquid ^{4}He is approximately $0.14\,\text{g/cm}^3$, this corresponds to an average distance between two atoms of approximately 4 Å, as the black circle in Fig. 20.8 left panel shows.

Now when ^{4}He is cooled one observes the variation in density shown in the right panel. At temperature above T_λ, the density increases as the temperature drops, but at T_λ the density is maximum and then decreases at lower densities. In other words, *the liquid expands when it becomes superfluid.*

According to the left panel in Fig. 20.8, if the liquid expands starting from an interatomic distance of 4 Å, it implies that the potential energy increases. The only way this can be possible is if the kinetic energy of the atoms decreases. This is direct experimental evidence that formation of the superfluid is associated with expansion of the fluid and associated with it a lowering of quantum kinetic energy.

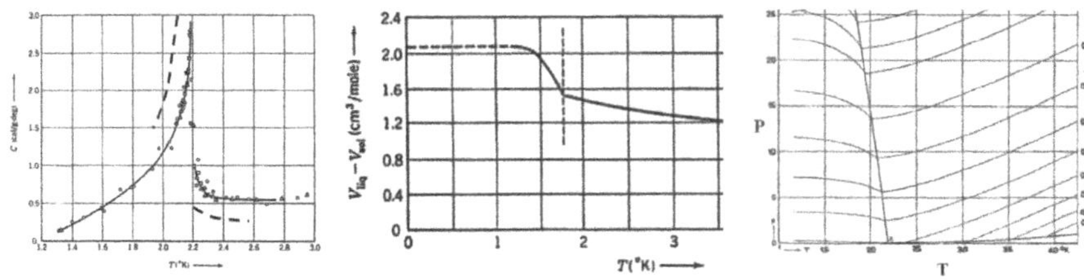

Fig. 20.9 Left panel: Specific heat of liquid ^{4}He as function of temperature; the dashed line indicates schematically the part of the specific heat associated with quantum kinetic energy, first term in Eq. (20.4). Center panel: volume difference between the liquid and solid states; right panel: pressure as function of temperature at constant density.

Figure 20.9 show results of three other experiments that say the same thing.

The left panel of Fig. 20.9 shows the specific heat, which is the derivative of energy with respect to temperature:

$$C = \frac{d\langle K\rangle}{dT} + \frac{d\langle U\rangle}{dT} \tag{20.4}$$

where $\langle K\rangle$ and $\langle U\rangle$ are average kinetic and potential energies. The second term is positive above T_λ, since the system expands as T increases, and the potential energy increases (Fig. 20.8). Below T_λ it is negative since the system expands when T decreases. Therefore, the first term in Eq. (20.4) is larger than the full line in the figure below T_λ and smaller above T_λ, as the dashed line indicates. Therefore, the jump at T_λ of the variation of kinetic energy with temperature is even bigger than that for the total energy, given by the full line. The fact that the kinetic energy varies (decreases) so rapidly below T_λ when the temperature decreases indicates that as the normal fluid condenses and becomes superfluid its quantum kinetic energy decreases substantially.

The central panel in Fig. 20.9 shows the change in volume between liquid and solid phases along the coexistence line between both phases. We will not go into the details at finite temperature, where we also have to consider the change in entropy that can contribute to an increase in volume when a system goes from solid to liquid. But at zero temperature the entropy doesn't change, and the fact that the volume of the liquid is larger than that of the solid at zero temperature indicates that in going from solid to superfluid the system expands because this lowers its kinetic energy.

Finally the right panel shows pressure versus temperature at constant density. Below T_λ the pressure *increases* when the temperature decreases. This indicates that when the superfluid forms it exerts additional quantum pressure, which explains why the fluid expands.

What is the significance of all this? That the transition of ^{4}He from normal fluid to superfluid is associated with expansion of the wavefunction, lowering of quantum kinetic energy, and increase in 'quantum pressure', which we can say originates in quantum zero point motion. This is precisely the same physics that we are saying governs the transition from a normal metal to the superconducting state, which we say is necessary to explain the Meissner effect, contrary to what BCS says.

As we had mentioned at the end of Chapter 2, already in 1911 Onnes discovered [9] that ^{4}He expands when the temperature is lowered, and he found this to be extremely interesting, so much so that he went back to investigating it in detail in 1922 [10], without however reaching any conclusion. The other effect that Onnes discovered in 1922 and is relevant to this discussion is what is called the "Onnes effect" [11], shown in the right panel of Fig. 20.10.

The right panel of Fig. 20.10 shows that superfluid in a jar spontaneously climbs up the walls and pours out, until there is no more fluid in the container. Incredible, isn't it? There have been a variety

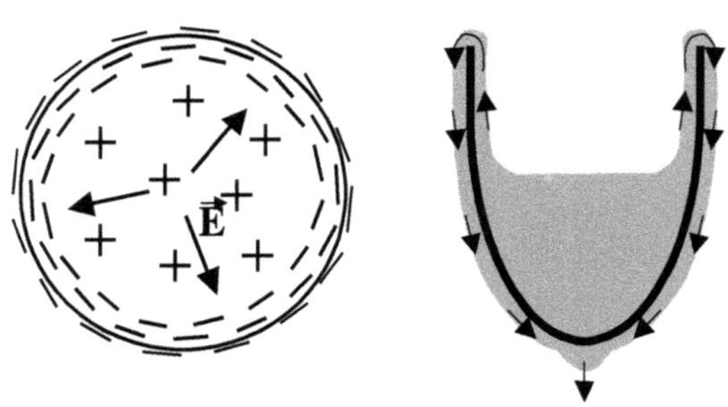

Fig. 20.10 Left panel: the superconductor expels electrons out of the solid. Right panel: the superfluid is expelled from the container out (Onnes effect).

of theories to explain this effect, we will not go into details because I don't agree with them. I believe it is a consequence of the same physics we are discussing. The superfluid expands, this pushes outward, and makes it flow out of the container.

The left panel of Fig. 20.10 shows the analogous in superconductors. We said that the wavefunction of electrons expands and electrons flow out. So much so (this we didn't mention before) that some electrons spill out beyond the surface, as Fig. 20.10 shows. Why don't they spill out completely, as superfluid helium on the right panel? Because the Coulomb attraction between electrons and the positive charge of ions prevents it, so they spill out a little only.

This intimate relation between superconductors and superfluid ^{4}He becomes even more apparent when we consider superfluid flow with zero potential difference. In superconductors this happens when we connect a superconducting wire between normal wires, as Fig. 20.11 left panel shows, and the same occurs in the 'double beaker' experiment of Daunt and Mendelssohn [12], designed specifically for this purpose. We show this in Fig. 20.11.

Kurt Mendelssohn, another German Jew exiled from Germany to Oxford in 1933, made many important experimental contributions to the understanding of liquid Helium. He pointed out [13, 14] the clear

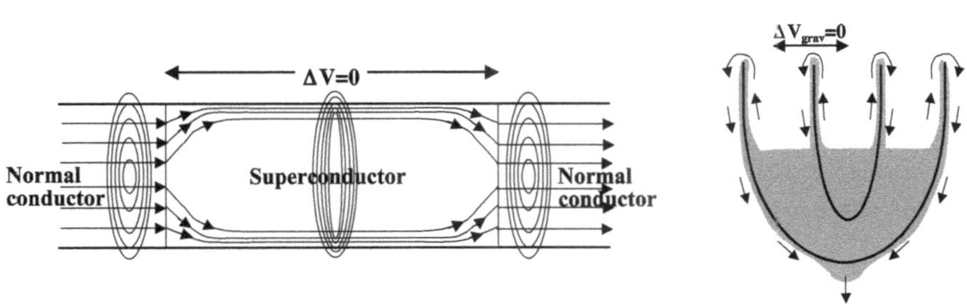

Fig. 20.11 Left panel: superconducting wire between two normal wires. There is no electric potential difference between both ends of the superconducting wire ($\Delta V = 0$), therefore there is no electric force on the charge carriers in this region. Right panel: 'double beaker' experiment of Mendelssohn. ^{4}He flows from the inner to the outer beaker and escapes from there, emptying both beakers. At all times, the level of the fluid in the inner and outer beaker is identical, therefore there is no gravitational potential difference between the surface of the fluids in the inner and outer beakers ($\Delta V_{\text{grav}} = 0$).

analogy between the phenomena shown in Fig. 20.11 and asked: what is the dynamical origin of these flows, that occur without any drop in potential, that is without any propelling force? He proposed that they were evidence for '*zero point motion*' of the condensed particles in the superfluid and the superconductor. He pointed out that "*neither case corresponds to a Bose–Einstein condensation since both have an appreciable zero-point energy*".

In addition, Daunt and Mendelssohn [15], Fritz London [16] who also did a lot of work on liquid Helium, and others [17], pointed out that the speed measured in ^{4}He for these films of superfluid that flow out of the container obeys the relation

$$v = \frac{\hbar}{2m_{\mathrm{He}}d} \tag{20.5}$$

where d is the thickness of the film, typically ~300 Å, yielding a velocity $v \sim 26\,\mathrm{cm/s}$. This relation can be interpreted as arising from Heisenberg's uncertainty principle $\Delta p \Delta x \sim \hbar/2$, with $\Delta p \sim m_{\mathrm{He}}v$ the uncertainty in the mechanical momentum and $\Delta x \sim d$ the uncertainty in position. Similarly we saw in Chapter 19 that the critical velocity for an electron in a superconductor is given by

$$v = \frac{\hbar}{4m_e\lambda_L} \tag{20.6}$$

that can also be interpreted as the velocity of an electron confined to a linear dimension $2\lambda_L$. Mendelssohn suggested [13] that these velocities as shown in Eqs. (20.5) and (20.6) are the velocities of 'zero point diffusion' of the particles in the condensate that originate in Heisenberg's uncertainty principle, and that this explains why the transport velocity is independent of external forces, i.e. occurs without any potential drop: the transport occurs because if at one end particles from the condensate are removed, zero point diffusion gives rise to flux in that direction. He emphasizes that "*the momentum of frictionless transport is not dissipated because it is zero-point energy*".

However, Mendelssohn's interpretation, even though it reveals very deep intuition, is not internally consistent. Heisenberg's uncertainty principle predicts that the momentum associated with spatial confinement should be in the same direction of the coordinate that is confined. Instead, both in the superfluid and superconductor the transport with speeds given by Eqs. (20.5) and (20.6) is parallel to the surface, i.e. perpendicular to the direction of confinement. It is clear that Heisenberg's uncertainty principle *is not* the explanation for superfluid film and superconducting current flow under zero potential difference. So what is it?

Superconductors give us the answer. The London–Mendelssohn transfer speed for superconductors Eq. (20.6) is the speed of electrons in orbits of radius $2\lambda_L$ in the absence of applied fields, that gives rise to the spin current, Eq. (19.9). The motion described by the speed Eq. (20.6) is *rotational* (Fig. 19.3). This means that both in superconductors and superfluid ^{4}He there must be *rotational zero point motion* in the ground state [18].

This is not part of BCS theory nor of any of the currently accepted theory to explain ^{4}He.

If the zero-point motion is rotational, it is easy to understand why spatial confinement in direction perpendicular to the surface gives rise to flow along the surface. It is rolling without slipping. Furthermore it is easy to understand the magnitude of the flow velocity, arising from quantization of angular momentum

$$\ell = mvd = \frac{\hbar}{2} \tag{20.7}$$

both for Eqs. (20.5) and (20.6). It is also easy to understand the origin of quantum pressure in these systems: the kinetic energy of rotational zero point motion decreases as the radius R of the rotational motion increases

$$E_{\text{kin}} = \frac{\ell^2}{2MR^2} \tag{20.8}$$

for particles of mass M in orbits of radius R with angular momentum ℓ. Thus, a particle rotating with fixed quantized angular momentum exerts quantum pressure to reduce its kinetic energy by expanding its orbit, and it does that in the transition to the superfluid or the superconducting state. The expanded orbits overlap, as for example

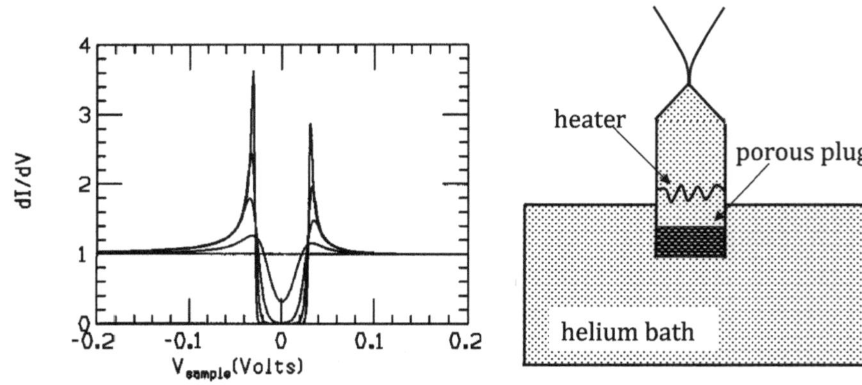

Fig. 20.12 Left panel: Conductance of a tunnel junction between superconductor and normal metal, showing that the current is larger when electrons come out of the superconductor (left peak) than when they come in (right peak). Right panel: fountain effect in superfluid ^{4}He, the lines on top represent fluid expelled upward. The heater reduces the concentration of superfluid in that region, hence the superfluid in the lower container exerts more pressure and flows upward with great force.

Fig. 19.3 shows, and this explains why it is necessary that there is 'phase coherence' to avoid collisions of particles in different orbits.

Finally, in Fig. 20.12, we show another example where this 'quantum pressure' is manifest in superconductors and in superfluid ^{4}He. As we saw in Chapter 16, in a tunnel junction the current is larger when the superconductor is negatively biased, reflecting the tendency of superconductors to expel electrons. For superfluid ^{4}He, this quantum pressure manifests itself vividly in the 'fountain effect' shown in the right panel of Fig. 20.12. When the liquid in the upper part is heated, the concentration of superfluid decreases, and the more concentrated superfluid in the colder region below moves up with great force, and is expelled through the top of the container. The porous plug in the figure allows for superfluid to move up but prevents backflow of normal fluid, which would suppress the effect.

In summary, we have seen in this chapter that both in superconductors and in ^{4}He the transition to the superconducting or superfluid states is associated with expansion of the wavefunction driven by lowering of quantum kinetic energy. It is reasonable that since both are macroscopic quantum systems they should have this physics in common, contrary to what BCS says. We have seen that several phenomena in superfluid ^{4}He and in superconductors are explained if the superfluids exert quantum pressure due to macroscopic quantum zero point motion. The frictionless flow of supercurrent in superconductors and the non-viscous flow of superfluid in ^{4}He and their respective critical velocities can be explained in a unified way as arising from quantum zero point diffusion. Both superconductors and ^{4}He expel superfluid outward because of quantum pressure, which manifests itself in the Meissner effect and tunneling asymmetry in superconductors, and in the Onnes effect and fountain effect in ^{4}He. Conventional theories have different explanations for each of these phenomena, the theory presented here has one unifying explanation. In Chapter 21, we will discuss another phenomenon that evidences this physics in an even clearer form.

References

[1] J. E. Hirsch, Electromotive forces and the Meissner effect puzzle, *J. Sup. Nov. Mag.* **23**, 309 (2010).

[2] J. E. Hirsch, Apparent violation of the conductivity sum rule in certain superconductors, *Physica C* **199**, 305 (1992).

[3] J. E. Hirsch, Superconductors that change color when they become superconducting, *Physica C* **201**, 347 (1992).

[4] H. J. A. Molegraaf, C. Presura, D. van der Marel, P. H. Kes, and M. Li, *Science* **295**, 2239 (2002).

[5] A. F. Santander-Syro, R. P. S. M. Lobo, N. Bontemps, Z. Konstantinovic, Z. Z. Li, and H. Raffy, *Europhys. Lett.* **62**, 568 (2003).

[6] J. E. Hirsch, The true colors of cuprates, *Science* **295**, 2226 (2002).

[7] J. E. Hirsch, Kinetic energy driven superfluidity and superconductivity and the origin of the Meissner effect, *Physica C* **493**, 18 (2013).

[8] D. Ceperley, *Rev. Mod. Phys.* **67**, 279 (1995).

[9] H. K. Onnes, Comm. N. *119 from the Physical Laboratory* at Leiden, 1911.

[10] H. K. Onnes and J. D. A. Boks, *Comm. N.* 170b, 1922.

[11] H. K. Onnes, *Leiden Comm.* 159; *Trans. Faraday Soc.* **18**(53) (1922).

[12] J. G. Daunt and K. Mendelssohn, *Nature* **157**, 839 (1946).

[13] K. Mendelssohn, *Proc. Phys. Soc. London* **57** (1945) 371.

[14] K. Mendelssohn, Report of an International Conference on Fundamental Particles and Low Temperatures, *Physical Society of London II*, 1947, p. 35.

[15] J. G. Daunt and K. Mendelssohn, Superconductivity and Liquid Helium II, *Nature* **150**, 604 (1942); J. G. Daunt and K. Mendelssohn, *Phys. Rev.* **69**, 126 (1946).

[16] F. London, *Rev. Mod. Phys.* **17**, 310 (1945).

[17] A. Bilj, J. De Boer, and A. Michels, *Physica* **8**, 655 (1941).

[18] J. E. Hirsch, *Mod. Phys. Lett.* **24**, 2201 (2010); *Phys. Lett. A* **374**, 3777 (2010).

Part V

ANSWERS TO THE KEY QUESTIONS AND CONCLUSIONS

Chapter 21

Spinning superconductors and ice skaters: the smoking gun

Even before the discovery of the Meissner effect, three German physicists, Becker, Heller, and Sauter, formulated [1] a question that gives very important clues on the nature of superconductivity — although nobody knows it. *What happens if we put a superconductor into rotation?*

Becker *et al.* didn't do an experiment. They did a theoretical calculation that told them that if they spin (put into rotation) a superconducting sphere, a magnetic field in the interior will be created that is parallel to the angular velocity of rotation, as Fig. 21.1 shows.

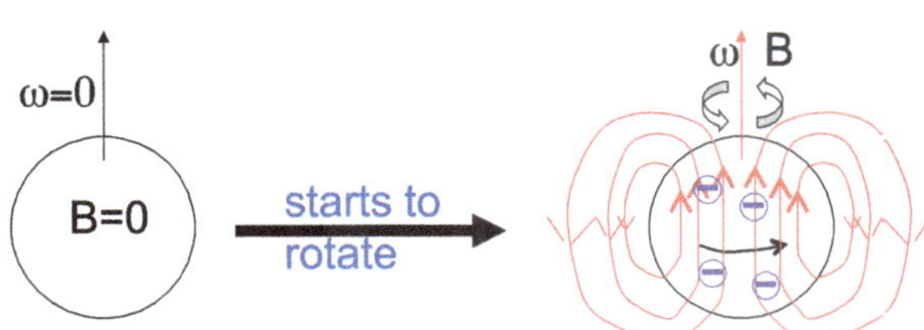

Fig. 21.1 Superconducting sphere initially at rest, then put into rotation. A magnetic field is generated in the interior (red field lines) that is uniform and parallel to the angular velocity of rotation, $\vec{\omega}$.

Becker *et al.* did the calculation only for a sphere, however the result is valid for bodies of any shape.

The first important clue that this result gives, assuming it is observed experimentally, is that superconductors 'know' what the

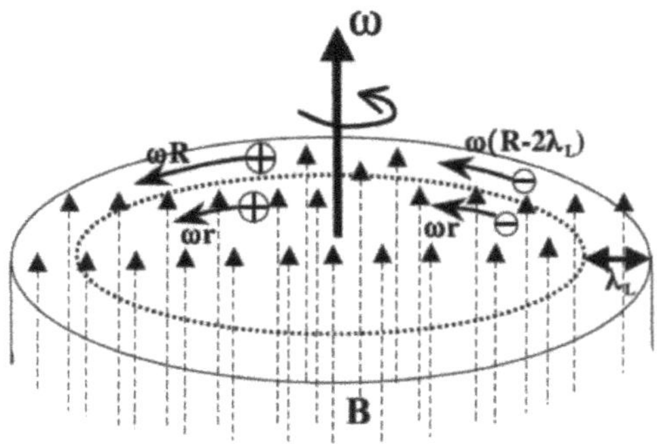

Fig. 21.2 Upper part of a rotating cylindrical superconductor. Electrons near the surface lag behind with respect to the ions, giving rise to a superficial electric current that generates a uniform magnetic field in the interior of the cylinder.

sign of the electric charge of the charge carriers is. That is what determines that the magnetic field is parallel, rather than antiparallel, to the rotation velocity. All superconductors show this orientation. Instead, it can be said that normal metals 'don't know' what the sign of the mobile charges is, since some show negative and some show positive Hall coefficients indicating that the mobile charge carriers are negative or positive depending on the metal.

Why is this magnetic field generated? Figure 21.2 shows in more detail what happens. From here on we focus on cylinders rather than spheres because both the physics and the mathematics are easier for that case. As Fig. 21.2 shows, electrons near the surface rotate a little slower than the ions, and for that reason an electric current near the surface exists that generates a uniform magnetic field $\vec{B}$ in the interior of the cylinder parallel to the angular velocity of rotation, $\vec{\omega}$.

Why does this happen? Qualitatively, they explained it as follows: in the superconductor the electrons move freely, they are detached from the ions. When the solid starts to rotate (assume in counterclockwise direction, as the figure shows) the ions move and the electrons initially don't. This generates a very large electric current in counterclockwise direction (the ions are positive) which generates a magnetic field parallel to the angular velocity $\vec{\omega}$ that increases with time, that in turn generates a Faraday electric field in clockwise direction, that pushes the electrons to follow the ions in counterclockwise direction at the same speed, except in a small layer adjacent to the surface where they lag behind a little, giving rise to the surface current.

Quantitatively, for a cylinder of radius R and angular velocity ω, the tangential speed of both ions and electrons in the interior at distance r of the cylinder axis is ωr, the speed of ions at the surface is ωR and the speed of electrons in the layer adjacent to the surface of thickness λ_L is $\omega(R - 2\lambda_L)$. From Eq. (6.13) of Chapter 6, inverting it, we obtain

$$B = -\frac{m_e c}{e\lambda_L} v_s \tag{21.1}$$

relating the magnetic field to the velocity of the superfluid near the surface. In this case, one has to use the relative velocity between electrons and ions, which is $\Delta v_s = 2\lambda_L \omega$. Replacing in Eq. (21.1) we obtain

$$B = -\frac{2m_e c}{e}\omega \tag{21.2}$$

as the magnetic field generated in the interior of a superconductor rotating with angular velocity ω. Quantitatively, the field is very small, $B = 7.144 \times 10^{-7} f(\text{Gauss})$, with $f = \omega/2\pi$ the number of revolutions per second.

It is not difficult to mathematically derive Eq. (21.2). The equation of motion for electrons in the superfluid is

$$m_e \frac{dv_s}{dt} = eE \tag{21.3}$$

assuming that the ions exert no force over the electrons when they begin to rotate (in the normal metal they do). The electric field E is determined by Faraday's law, Eq. (6.1b), that for cylindrical symmetry gives

$$E(r,t) = -\frac{1}{2\pi r}\frac{1}{c}\frac{d\phi(r,t)}{dt} \tag{21.4}$$

at distance r from the cylinder axis. The magnetic flux in the interior of r is, assuming a uniform magnetic field, $\phi(r,t) = \pi r^2 B$. Replacing in Eq. (21.3),

$$m_e \frac{dv_s}{dt} = eE = -\frac{e}{2c} r \frac{dB.}{dt}. \tag{21.5}$$

In the interior, electrons and ions move at the same speed, so that $v_s = \omega r$. Using this and integrating Eq. (21.5) in time, Eq. (21.2) results. Adjacent to the surface, instead of $v_s = \omega r$ we have

$$v_s = \omega(R - 2\lambda_L) = \omega R - 2\lambda_L \omega = \omega R + \Delta v_s \tag{21.6}$$

as we said before, and we already showed that this velocity Δv_s gives rise to the magnetic field Eq. (21.2). Everything is consistent.

Up to here, no problem. Fortunately, Becker *et al.* didn't think of asking the following question. Assume we rotate a *normal metal* with angular velocity $\vec{\omega}$. Electrons and ions rotate together, there is no electric current nor magnetic field. What happens if we cool down while the metal is rotating and it becomes superconducting?

If Becker *et al.* had asked themselves that question, they would have answered: nothing happens! Inertia causes electrons and ions to continue rotating together, since there is no difference in the speeds of electrons and ions, there is no electric field and no magnetic field is generated in the rotating metal when it becomes superconducting.

They would have been wrong.

Shortly after the Becker *et al.* article, Meissner discovered his effect, and the London brothers developed their theory. In the London theory, the state of the superconductor in a magnetic field *is independent of history.* If we apply a magnetic field to a superconductor, it doesn't let it go in. If we cool a normal metal in a magnetic field, the superconductor expels it. The final state is the same in both cases: there is no magnetic field in the interior of a superconductor at rest.

In the same way, reasoned Fritz London [2], the state of a rotating superconductor cannot depend on history. If we cool a rotating normal metal, the same magnetic field Eq. (21.2) has to be generated as the one that is generated according to the calculation of Becker *et al.* when we put a superconductor into rotation.

This means that when the rotating normal metal becomes superconducting, electrons near the surface have to spontaneously slow down a little to acquire the relative velocity Δv_s that generates the magnetic field in the interior, defying the law of inertia.

This seemed "quite absurd" to Fritz [3], intuitively, but he had no doubt that it had to be that way according to his theory. And indeed, when many years later the experiment was done, what London expected was measured [4–6] (unfortunately London had already passed away). The same magnetic field Eq. (21.2), no matter whether a superconductor at rest was put into rotation, or a normal metal in rotation was cooled.

Because of this, the magnetic field of rotating superconductors Eq. (21.2) is called the 'London field' in the literature, it is not called the 'Becker field'. I would have called it the 'Becker–London field' to be fair to both.

But now we need to explain: how do electrons spontaneously slow down when we cool a rotating normal metal? Doesn't the law of inertia say they shouldn't?

BCS says: "they slow down because the equations say that the final state has to have a magnetic field, Eq. (21.2). The London Eq. (7.4), with $\vec{v}_s = \vec{\omega} \times \vec{r}$ and $\vec{\nabla} \times (\vec{\omega} \times \vec{r}) = 2\vec{\omega}$, yields Eq. (21.2). In order to generate that field, electrons near the surface have to slow down. How they do it, what is the process, we don't know but it doesn't matter. Quantum mechanics somehow explains it."

I say, no. According to Bohr's correspondence principle, it has to be understandable using laws that are familiar to us in the macroscopic realm. And I say the explanation is very simple [7], as Figs. 21.3 and 21.4 show.

If something is rotating, in order to slow it down, we need to increase its 'moment of inertia'. The mathematical relation is $L = I\omega$ with L the angular momentum, I the moment of inertia, and ω the angular velocity, analogous to the relation $p = mv$ between linear momentum, mass, and linear velocity. L is constant, if ω decreases it is because I increases. For the ice skater of Fig. 21.3, her moment of inertia increases when she extends her arms. The moment of inertia I of a body is proportional to MR^2, where M is the mass of the body and R is an average of the extension of the body in direction perpendicular to the angular velocity of rotation $\vec{\omega}$. When a body *expands*

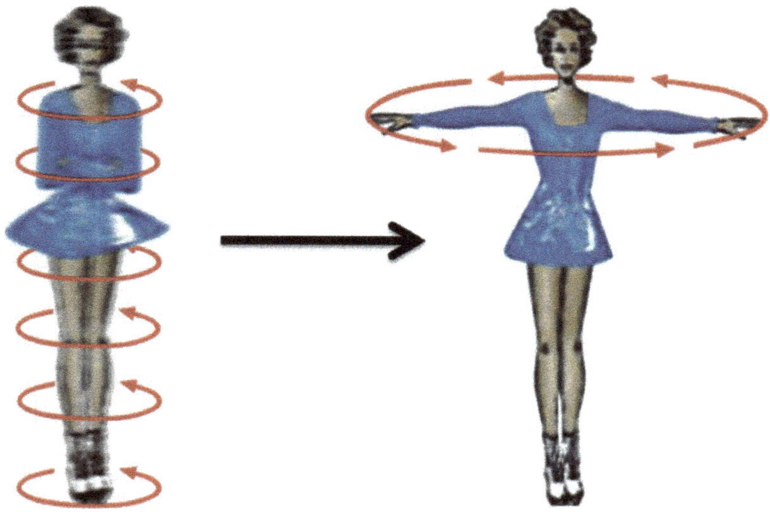

Fig. 21.3 Ice skater spinning rapidly (left panel). If she extends her arms, her moment of inertia increases and the rotation velocity decreases (right panel).

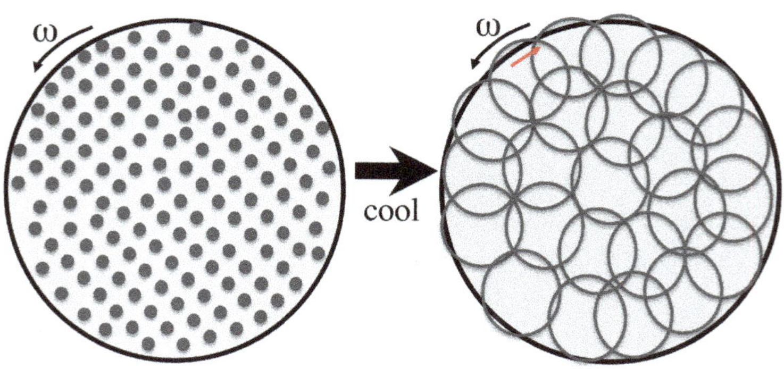

Fig. 21.4 Spinning normal metal (left), the little circles are electrons in microscopic orbits of radius k_F^{-1}. When it is cooled into the superconducting state (right), the electronic orbits expand, their moment of inertia increases, and the speed of rotation of electrons (near the surface) decreases.

in direction perpendicular to $\vec{\omega}$, R increases and as a consequence its moment of inertia increases and it slows down, rotates more slowly.

So, if the electrons that were rotating at the same speed as the body in the normal state rotate slower when they become superconducting, it means their moment of inertia increased, they *expanded.*

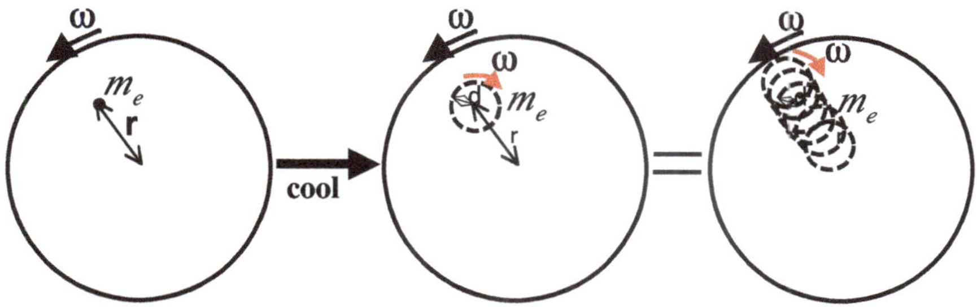

Fig. 21.5 Expansion of the electron, from a point orbit to an orbit of radius d, in a rotating cylinder. To conserve angular momentum, the electron acquires angular velocity ω in its orbit in opposite direction to the rotation of the cylinder (red arrow). When many orbits are superposed, there is only relative velocity between electrons and the cylinder in the layer of thickness λ_L adjacent to the surface, the velocities in the interior cancel out.

Haven't we talked about this before? Yes, in Chapters 19 and 20 and in Fig. 20.3. We said that in the transition to superconductivity electrons expand their orbits from microscopic radius (k_F^{-1}) to mesoscopic radius $2\lambda_L$. That is what Fig. 21.4 shows qualitatively.

Let us now see how this explains the London field quantitatively. Figure 21.5 shows in more detail what happens when the electronic orbit expands in the rotating superconductor.

On the left panel, we have a rotating normal metal, the electrons in orbits of radius k_F^{-1} we may consider essentially to be point particles of mass m_e. For an electron at distance r from the cylinder axis, its moment of inertia is $i_e = m_e r^2$. Suppose the electron expands to a ring of radius d centered at the same position r. Its moment of inertia with respect to the axis of the cylinder increases as the following equation shows:

$$i_e = m_e r^2 \rightarrow m_e(r^2 + d^2). \tag{21.7}$$

The electron's angular momentum ℓ is given by $\ell_e = i_e\omega = m_e r^2\omega$, so it would also increase:

$$\ell_e = m_e r^2\omega \rightarrow m_e(r^2 + d^2)\omega. \tag{21.8}$$

But if the electron is completely free, uncoupled from the solid, it has to conserve its angular momentum. So what happens is that this ring of radius d as it expands starts to rotate in clockwise direction, contrary to the rotation of the cylinder, with the same angular velocity ω, and its total angular momentum is

$$\ell_e = m_e r^2\omega \rightarrow m_e(r^2 + d^2)\omega - m_e d^2\omega = m_e r^2\omega \tag{21.9}$$

that is, it didn't change!

So we can think that all the electrons in the superfluid expanded to orbits of radius $d = 2\lambda_L$, each increasing its moment of inertia by

$$\Delta i_e = m_e(2\lambda_L)^2 \tag{21.10}$$

and in the process of expanding they also acquired an angular velocity ω in opposite direction to the cylinder rotation. As we already saw in Fig. 19.1, superposition of these orbits cancels the internal motion as the right panel of Fig. 21.5 indicates, what survives is the motion of the electrons in the layer of thickness λ_L adjacent to the surface, with relative velocity with respect to the ions $\Delta v_s = 2\lambda_L\omega$, that is the radius of the expanded orbit multiplied by the angular velocity in clockwise direction (red arrow in Fig. 21.5).

That is precisely the relative velocity of electrons in the surface layer necessary to generate the London field, as we saw earlier.

To convince ourselves, let us redo the analysis, now globally for a cylinder of radius R, height h, with superfluid density n_s, rotating with angular velocity ω. To generate the London field Eq. (21.2), it is necessary that electrons in the layer of thickness λ_L adjacent to the surface reduce their speed by $\Delta v_s = 2\lambda_L\omega$. This gives rise to a decrease in the angular momentum of electrons near the surface (see Eq. (19.1))

$$\Delta L_e^{\text{sup}} = -n_s(2\pi R\lambda_L h)[m_e\Delta v_s R) = -m_e(2\lambda_L)^2 n_s(\pi R^2 h)\omega. \tag{21.11}$$

On the other hand, in the interior of the cylinder there is no current. The angular momentum of electrons in the interior, $L_e^{\text{int}} = I_e\omega$, will change if there is a change in the moment of inertia:

$$\Delta L_e^{\text{int}} = (\Delta I_e)\omega \tag{21.12}$$

and conservation of angular momentum tells us that it has to compensate the reduction of angular momentum of the surface layer, that is:

$$\Delta L_e^{\text{sup}} + \Delta L_e^{\text{int}} = 0 \tag{21.13}$$

from which we deduce that the moment of inertia of all the electrons in the cylinder increased by

$$\Delta I_e = [n_s(\pi R^2 h)](m_e(2\lambda_L)^2). \tag{21.14}$$

The factor in square brackets is the number of electrons, the factor that follows is the change in the moment of inertia of each electron, coincident with Eq. (21.10).

For even more evidence, consider the case of a rotating *cylindrical shell* as shown in Fig. 21.6. We assume that the thickness of the shell is much larger than the London penetration depth λ_L.

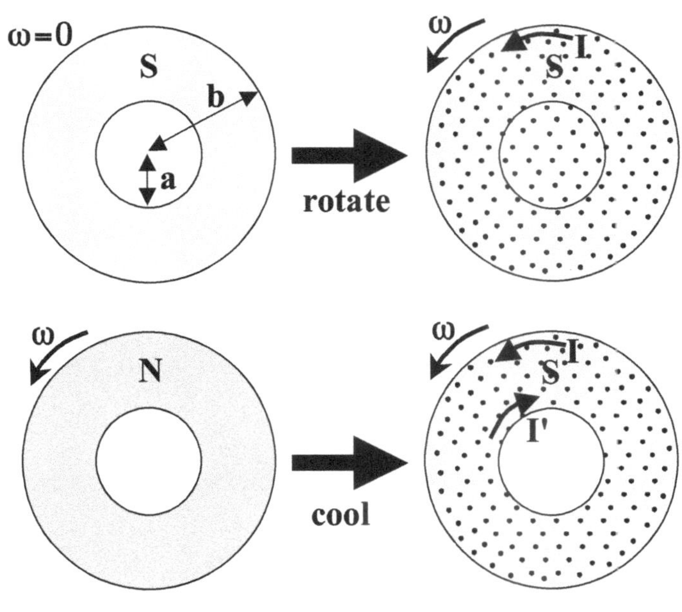

Fig. 21.6 Behavior of a rotating hollow cylinder. Magnetic field coming out of the paper is indicated by dots. If the cylinder is superconducting at rest and is set into rotation, a magnetic field both in the shell and in the hollow interior is generated (upper panels). Instead, if the normal metal is rotating and is cooled into the superconducting state, a magnetic field in the material is generated but not in the hollow interior (lower panels).

Here two different behaviors are observed, as the figure shows. If the cylindrical shell is initially superconducting and starts to rotate, only a current near the outer surface is generated, which generates a magnetic field both in the region of the shell and in the hollow interior. Instead, if the rotating normal metal is cooled, a magnetic field is only generated in the region of the shell [6]. The latter means that two currents are generated, one at the outer surface and an equal one in opposite direction at the inner surface of the shell. How are these currents generated?

Figure 21.7 shows what the electrons do. Electrons near the outer surface slow down, just like in the case of the solid cylinder that we discussed earlier. Electrons near the inner surface need to speed up, in order to cancel the magnetic field in the interior.

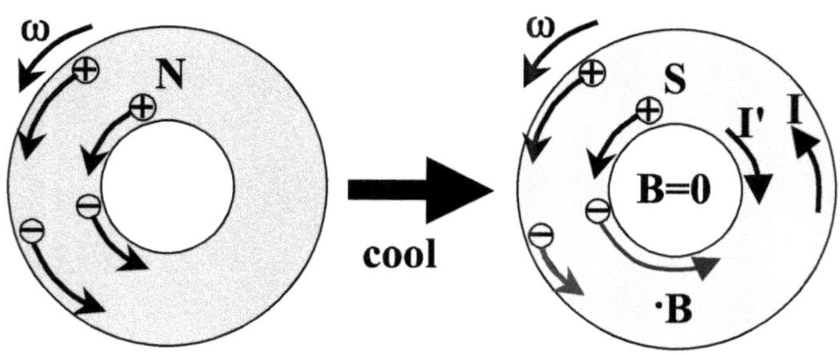

Fig. 21.7 In the normal state of the cylindrical shell in rotation, electrons and ions rotate with the same velocity. When the metal is cooled and enters the superconducting state, electrons near the outer surface slow down and those near the inner surface speed up.

Does it look more difficult to explain this more complicated behavior? No, it is equally simple, as Fig. 21.8 shows.

Exactly the same process that we explained earlier. The electron expands its orbit to radius $2\lambda_L$, in the process it acquires angular velocity in clockwise direction. The velocities compensate in the interior of the shell and survive near both surfaces, at the outer surface they go in clockwise direction and at the inner surface in counterclockwise direction, which results in electrons slowing down near the outer surface and speeding up near the inner surface.

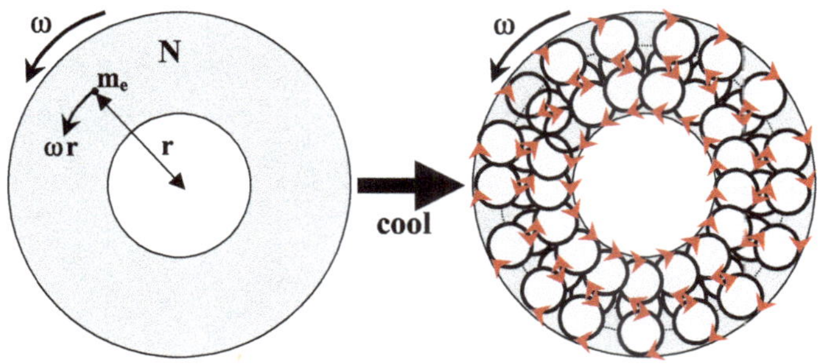

Fig. 21.8 When the rotating shell enters the superconducting state, the electronic orbits expand and start to rotate in direction opposite to the shell rotation. The red arrows show how this accelerates the electrons near the outer surface and slows down the electrons near the inner surface.

The mathematics that describes this is a simple extension of what we saw before. To generate the magnetic field in the shell but not in the interior, electrons at the outer surface slow down by $\Delta v_s = 2\lambda_L\omega$, and electrons at the inner surface speed up by the same amount. Let us call a and b the radii of the inner and outer surfaces respectively. The total change in angular momentum of electrons near the two surfaces is, generalizing Eq. (21.11)

$$\begin{aligned}\Delta L_e^{\text{sup}} &= -n_s(2\pi b\lambda_L h)[m_e\Delta v_s b) + n_s(2\pi a\lambda_L h)[m_e\Delta v_s a) \\ &= -m_e(2\lambda_L)^2 n_s(\pi(b^2-a^2)h)\omega. \qquad (21.15)\end{aligned}$$

In order for the total change in angular momentum to be zero, as Eq. (21.13) says, the change in moment of inertia has to be, instead of Eq. (21.14)

$$\Delta I_e = [n_s(\pi(b^2-a^2)^2 h)](m_e(2\lambda_L)^2). \qquad (21.16)$$

which again is simply the number of electrons in the shell (factor in square brackets) times the change in moment of inertia of each electron, Eq. (21.10).

To conclude, let us see what happens if we put a cylinder of a type II superconductor into rotation. We will not go into details of explaining what a type II superconductor is (most superconducting compounds are type II), let's just point out that when a magnetic

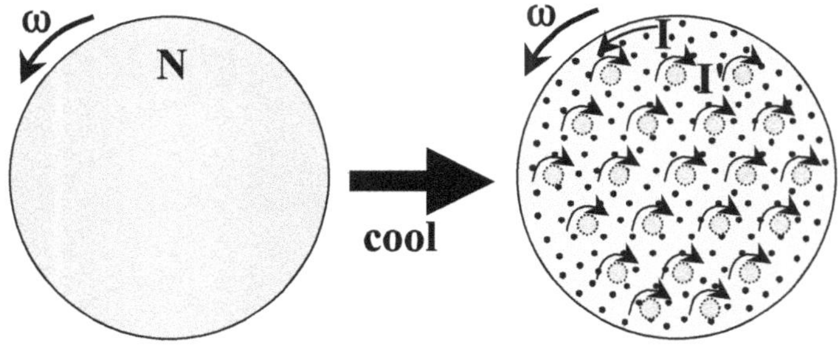

Fig. 21.9 When a rotating normal cylinder enters the mixed state of type II superconductors, a very complicated pattern of currents develops, indicated by the arrows in the figure.

field is applied there is a regime called 'mixed state' where the magnetic field penetrates in the form of tubes called vortices, where the system is normal inside the vortex and superconducting outside.

Figure 21.9 shows what happens if we cool a rotating cylinder and it enters the mixed state. A pattern of currents indicated by the arrows in the figure is generated. That is, near the outer surface electrons slow down, and around the little tubes, where the system remains in the normal state, they speed up. It is easy to understand how expansion of orbits in the superconducting part, conserving their angular momentum, will give rise to this pattern of velocities, analogously to Fig. 21.7. It is difficult to imagine how such a complicated current pattern could be generated by any other mechanism.

In summary, rotating superconductors show us in a crystal clear way that the transition to the superconducting state is associated with expansion of orbits. We had already seen this in earlier chapters in connection with the Meissner effect, but here it is much clearer. We don't have to know anything about Faraday's law nor electromagnetism. What other explanation could there be for something that is rotating to spontaneously slow down that is *not* increase in the moment of inertia due to expansion, like the ice skater in Fig. 21.3 or the electrons in Fig. 21.4? And the fact that for more complicated cases, like the cylindrical shell or type II superconductors, the same concepts explain the complicated patterns of currents that get generated, strongly suggests that it is the correct explanation. It is for this reason that we call rotating superconductors 'the smoking gun' that reveals this fundamental physics of superconductors, not described by BCS theory.

It is interesting to note that practically none of the contemporary textbooks on superconductivity mention rotating superconductors. Surely it is because there is no explanation for their behavior within BCS theory, so better not to mention them. For that reason, many physicists working on superconductivity today don't even know that rotating superconductors have a magnetic field in their interior. It should also be pointed out that this behavior of rotating superconductors has been experimentally verified both for so-called 'conventional' as well as for 'unconventional' superconductors, including the cuprates.

References

[1] R. Becker, G. Heller, und F. Sauter, Über die Stromverteilung in einer supraleitenden Kugel, *Zeitschrift für Physik* **85**, 772 (1933).
[2] F. London, *Superfluids*, Vol. I, Dover, New York, 1961.
[3] F. London, Ref. [2], p. 82.
[4] A. F. Hildebrandt, Magnetic field of a rotating superconductor, *Phys. Rev. Lett.* **12**, 190 (1964).
[5] N. F. Brickman, Rotating superconductors, *Phys. Rev.* **184**, 460 (1969).
[6] J. B. Hendricks, C. A. King, and H. E. Roschach, Magnetization by rotation: The Barnett effect in a superconductor, *J. Low. Temp. Phys* **4**, 209 (1971).
[7] J. E. Hirsch, Moment of inertia of superconductors, *Phys. Lett. A* **383**, 83 (2019); *Ann. Phys.* **531**, 1900212 (2019).

Chapter 22

The Meissner effect explained (almost all)

Let us now return to the Meissner effect and understand it in detail, including conservation of momentum. Consider Fig. 9.1, lower panel. We want to explain how the superconducting region expands from the center to the surface of a cylinder that is becoming superconducting in the presence of a magnetic field, and how in the process the magnetic field is expelled. Figure 22.1 shows what has to be explained.

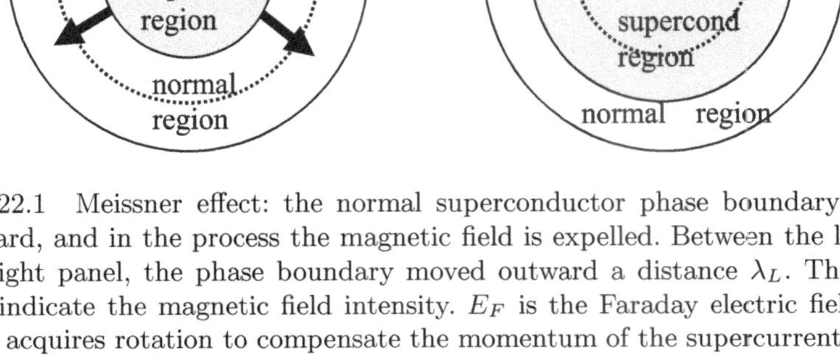

Fig. 22.1 Meissner effect: the normal superconductor phase boundary moves outward, and in the process the magnetic field is expelled. Between the left and the right panel, the phase boundary moved outward a distance λ_L. The black dots indicate the magnetic field intensity. E_F is the Faraday electric field. The body acquires rotation to compensate the momentum of the supercurrent, i.e. to conserve angular momentum.

The figure shows how the phase boundary between the superconducting and normal phases moves out expelling an applied magnetic field, H_c. H_c is the 'critical' magnetic field for the given temperature, that exists at the phase boundary between superconducting and normal phases. Between the left and right panels the phase boundary moved outward a distance λ_L. Let us examine what happened.

(1) The green electrons in the left panel are part of the supercurrent close to the surface of the superconducting region that nullifies the magnetic field in the interior. In the right panel, those electrons are already far from the surface inside the superconducting region, so they don't carry supercurrent (recall that in superconductors supercurrent is only carried in a layer of thickness λ_L adjacent to the surface, or adjacent to the phase boundary between superconducting and normal phases). How did those electrons stop their motion?

(2) The red electrons on the left panel are in the normal phase, so they don't transport supercurrent. In the right panel, they are in the superconducting phase, in the layer of thickness λ_L adjacent to the phase boundary, so they transport supercurrent with the velocity v_s given by Eq. (6.13), with $B = H_c$. How did those electrons acquire that velocity?

(3) The electronic angular momentum in the right panel is larger than in the left panel, because there are more electrons transporting supercurrent, and because they are farther from the origin (we will see quantitatively how much larger in a moment). Therefore, the body must have acquired clockwise momentum (indicated as 'body rotation' in the figure). How did the body acquire that momentum?

We also need to take into account that the process is thermodynamically *reversible*. Which means, the explanation that we find for these processes has to work equally well for the process in opposite direction, that is from right to left, inverting the arrows that need to be inverted.

The Faraday electric field E_F appears because of the change in magnetic flux, and is in counterclockwise direction, i.e. it tries to create a counterclockwise current to prevent the change in magnetic flux, i.e. to prevent the flux expulsion. So:

(1) E_F pushes the green electrons in clockwise direction (due to the negative charge of the electron), so it tends to slow down the green electrons, and is in fact responsible for the fact that in the right panel the green electrons no longer transport supercurrent. No problem here.

(2) E_F also pushes the red electrons in clockwise direction, but they ignore it and instead start moving in counterclockwise direction. What is the force that makes them move despite the opposition of E_F?

(3) E_F pushes the ions in counterclockwise direction. But the positive ions together with the body acquire clockwise momentum. What is the force that makes the ions move despite the opposition of E_F?

The answer, qualitatively, is in Fig. 22.2, left panel.

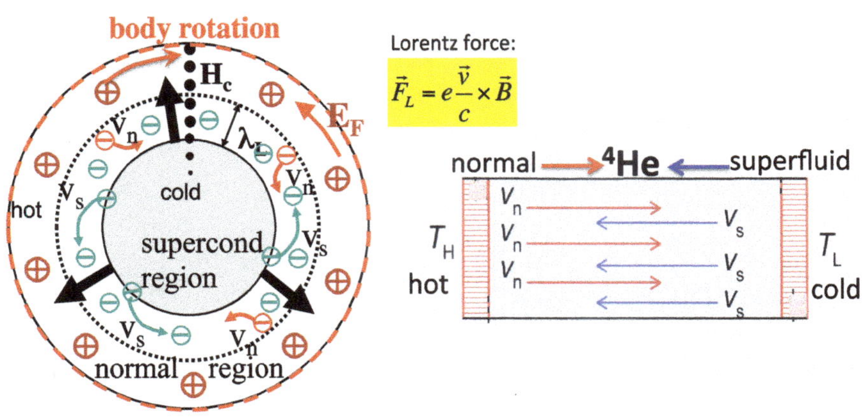

Fig. 22.2 Left panel: dynamics of the Meissner effect. The boundary between superconducting region (inside) and normal region (outside) moves outward. There is an outflow of electrons becoming superconducting (green) and a backflow (inflow) of normal electrons (red). The result of these radial flows affected by the Lorentz force F_L is that superconducting electrons acquire azimuthal momentum in counterclockwise direction and the body acquires azimuthal momentum in clockwise direction, satisfying momentum conservation. The right panel shows analogously a process of flow of superfluid and backflow of normal fluid for liquid helium.

The electrons that acquire counterclockwise momentum and become part of the supercurrent are electrons that, in the process of going from normal to superconducting, move radially out a distance λ_L, and in that process acquire the speed of the Meissner current, Eq. (6.13), as shown by the green electrons in the figure.

The algebra is simple. Under the action of the Lorentz force F_L indicated in the figure, with $B = H_c$, the momentum in direction parallel to the phase boundary (azimuthal) acquired by the electron moving radially out a distance Δr in time interval Δt is

$$\Delta p_{//} = -\frac{e}{c}\int v_r B dt = -\frac{e}{c}\Delta r B \tag{22.1}$$

where $v_r = \Delta r/\Delta t$ is the speed of radial motion. Assuming $\Delta r = \lambda_L$, we obtain, with $p_{//} \equiv m_e v_s$, from Eq. (22.1)

$$v_s = -\frac{e\lambda_L}{m_e c}B \tag{22.2}$$

i.e. the velocity of electrons in the Meissner current, Eq. (6.13).

What happens with the force exerted by the Faraday field, that acts in the opposite direction? Let's see. From Faraday's law, we deduce that the Faraday field is given by

$$E_F = \frac{\dot{r}_0}{c}H_c \tag{22.3}$$

where $\dot{r}_0$ is the speed at which the phase boundary is moving outward. The magnitude of $\dot{r}_0$ depends on the resistivity of the normal metal, it can be calculated and measured, and is of order cm/s [1], we will not discuss that here. Taking into account the force due to the Faraday field, Eq. (22.1) is modified to

$$\Delta p_{//} = -\frac{e}{c}\Delta r B + eE_F\Delta t \tag{22.4}$$

therefore, we can ignore the second term if the condition

$$\dot{r}_0 \ll v_r \tag{22.5}$$

is met, that is, if the speed of expulsion of the electron outward is much larger than the speed of motion of the phase boundary.

In reality, what happens is that electrons at the phase boundary going from normal to superconducting expand their orbits to radius $(2\lambda_L)$, and as we saw in Fig. 19.2, this causes the electron to acquire azimuthal speed Eq. (22.2), so it is equivalent to think that the electron moved out a distance λ_L.

This process generates a density of *radial* current of electrons that are becoming superconducting, of magnitude

$$J_r = en_s\dot{r}_0 \tag{22.6}$$

and creates a radial electric field that causes a backflow of normal electrons in radial direction inward. These electrons (red in Fig. 22.2) are also deflected by the Lorentz force, in clockwise direction, as the figure shows, acquiring azimuthal momentum in clockwise direction. These electrons, being normal rather than superconducting, transfer their momentum to the body, making it rotate in clockwise direction, as the figure shows. This is how conservation of momentum is satisfied. We had already mentioned this backflow in Fig. 18.2.

These processes, of radial outflow of superconducting electrons and radial inflow of normal electrons, are completely analogous to what occurs in liquid helium in the fountain effect discussed in Chapter 20. The right panel of Fig. 22.2 shows those processes in helium.

It remains to be explained how exactly do the normal backflowing electrons transfer their azimuthal momentum acquired through the Lorentz force to the body, to make it rotate and thus conserve momentum. In 2015, I wrote a paper [2] explaining the dynamics of the Meissner effect as I explained it above, saying that these normal backflowing electrons collide with impurities in the body, and in that way transmit their azimuthal momentum to the solid.

A referee of that paper asked me: what role do holes, that you have been talking about for so many years, play in this process? I answered: *"In many other papers the key issue of holes versus electrons is addressed in detail, but it is not relevant to this paper and there is no reason to address it here."* And in the paper I said verbatim "*As the phase boundary advances at rate* $\dot{r}_0$, *normal (negatively charged) carriers in the boundary layer are backflowing at speed* $\dot{r}_0$, *or equivalently positive normal carriers (holes) move forward together with the phase boundary. Because normal carriers scatter off the lattice they do not acquire a large azimuthal speed from the action of the magnetic Lorentz force; instead they transfer their azimuthal momentum to the lattice as a whole, thus accounting for momentum conservation.*"

It was wrong, in particular the word "*equivalently*". There is no equivalency! The following year I had to write a Corrigendum to that paper [3], where I included an apology to the referee for having dismissed his/her question.

The reason why what I said above was wrong is because processes where electrons collide with impurities are not reversible processes, they generate entropy. But as we said earlier, the metal-superconductor phase transformation is thermodynamically reversible, and in reversible processes no entropy is generated. In Chapter 23, we will see how holes resolve this difficulty.

But before this, let us go back to Fig. 9.1, lower panel. BCS agrees that the superconducting phase expands, but it doesn't say that charge is expelled. How does it say that it expands? The region where 'phase coherence' exists expands, because the superconducting region has it and the normal region does not. But BCS doesn't explain what does mean. By contrast, in our description the phase coherence is explained naturally from the expansion of the orbits that occurs when electrons enter the superconducting state. Expanded orbits necessarily overlap, and it is necessary that electrons moving in those orbits have their 'phases' (interpreted as the position of the electron along the orbit) coherent, i.e. all at the same angular position at every instant, to avoid collisions between different electrons. Figure 22.3 shows this schematically. Instead, in the normal phase, the orbits of radius k_F^{-1} do not overlap, hence it is not necessary that phase coherence exists, and in fact it doesn't.

Figure 22.3 shows that in our theory there is a direct and logical connection between (1) expansion of the superconducting phase, (2) establishment of phase coherence, and (3) how electrons acquire

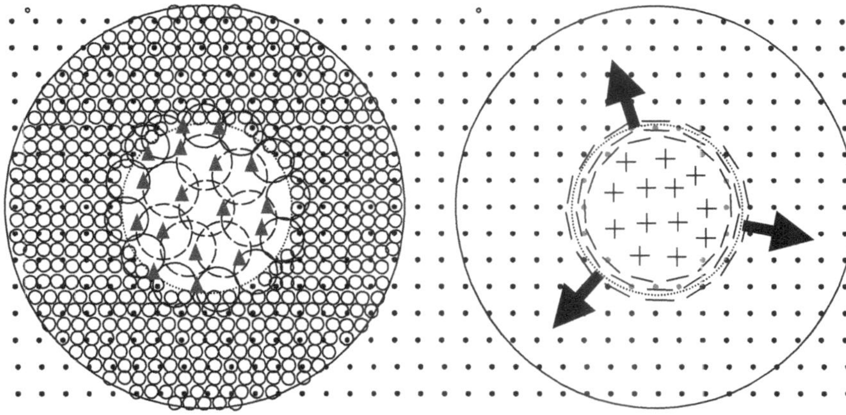

Fig. 22.3 Left panel: the triangles show the "phase", understood as the position of the electrons along their expanded orbits in the superconducting region. Because the orbits of radius $2\lambda_L$ are overlapping there has to be 'phase coherence', because if the phases are arbitrary there would be collisions between electrons. In the normal phase orbits do not overlap, and for that reason there is no phase coherence. The right panel shows the distribution of charge, more negative charge at the superconductor-normal phase boundary than in the interior.

the velocity of the supercurrent. The three things are explained by the same, *expansion of the orbits*. In contrast, BCS theory says that (1), (2), and (3) happen, but it does not explain how each one of them happens nor what is the relation between them.

References

[1] J. E. Hirsch, Dynamics of the normal-superconductor phase transition and the puzzle of the Meissner effect, *Annals of Physics* **362**, 1 (2015).

[2] J. E. Hirsch, On the dynamics of the Meissner effect, *Physica Scripta* **91**, 035801 (2016).

[3] J. E. Hirsch, Corrigendum: On the dynamics of the Meissner effect (2016 *Phys. Scr.* 91 035801), *Physica Scripta* **91**, 099501 (2016).

Chapter 23

The secret of the holes

After having worked on what I call the "theory of hole superconductivity" for 28 years, finally arriving at an understanding of the Meissner effect, in the homestretch so to speak, I forgot about holes when I wrote the paper mentioned in 2015, Ref. 2 of Chapter 22.

Ten years earlier in a paper [1] I had given a long list of 18 properties that are different for electrons and holes that I proposed are relevant to superconductivity, reproduced in Fig. 23.1. Right in the middle I had listed the most important one, not quite knowing why, i.e. not understanding what it has to do with superconductivity. I didn't realize why it is so important until 11 years later [2–4]. Fortunately, or unfortunately, nobody else realized it either. Let's see.

Let us go back to Chapter 5, Fig. 5.2, Hall effect. I claim that it gives rise to the following 'mystery', illustrated in Fig. 23.2. The magnetic Ampere force acts on conductors carrying an electric current when there is an applied magnetic field, according to the formula given in the figure, that I reproduce here:

$$\vec{F}_{\rm Amp} = \frac{I}{c}\vec{L} \times \vec{H}. \qquad (23.1)$$

L is the length of the conductor, I is the current that circulates, that is proportional to J_x in the figure, the current density. The Ampere force including its direction is the same independent of the Hall coefficient.

Bonding electron at the Fermi energy	Antibonding electron at the Fermi energy
Undressed	Dressed
Low kinetic energy	High kinetic energy
Long wavelength	Short wavelength
Small effective mass	Large effective mass
Uniform charge density	Nonuniform charge density
Moves in direction of force	Moves opposite to force
Conducts electricity	Anticonducts electricity
Contributes to Drude weight	Anticontributes to Drude weight
Detached from lattice	Transfers momentum to lattice
Large quasiparticle weight	Small quasiparticle weight
Coherent conduction	Incoherent conduction
Large Drude weight	Small Drude weight
Negative Hall coefficient	Positive Hall coefficient
Good metals	Bad metals
Stable lattices	Unstable lattices
Ions attract each other	Ions repel each other
Carriers repel each other	Carriers attract each other
Normal metals	Superconductors

Fig. 23.1 Differences between electrons (left column) and holes (right column) that are relevant to understand superconductivity [1].

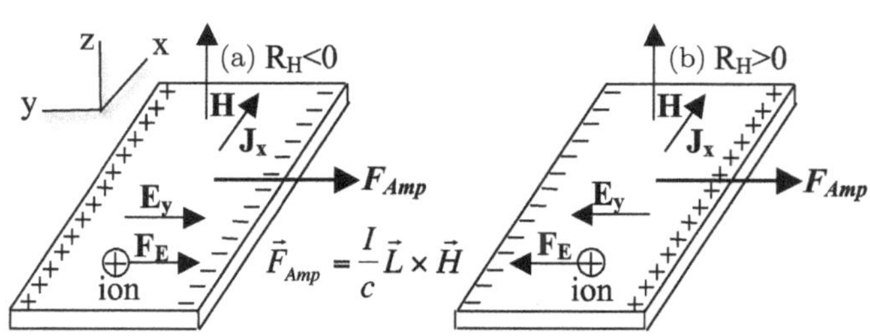

Fig. 23.2 Hall bars for a metal of negative Hall coefficient (left) where the charge carriers are electrons, and a metal with positive Hall coefficient (right) where the charge carriers are holes. L is the length of the bars, I is the current. The Ampere force $F_{\rm Amp}$ is the same in both cases. How is that possible?.

On the left panel of the figure, the Ampere force is easily understood. As will be remembered from Chapter 5, electrons are deflected to the right edge of the bar, creating an electric field E_y that goes from left to right. This field E_y exerts a force on the positive ions that points to the right, F_E. This force acting on the ions is precisely the Ampere force. No problem.

The problem is the right panel. Here, the transverse polarity is opposite to the left panel because the Hall coefficient R_H is positive instead of negative. As will be remembered from Chapter 5, Fig. 5.2, this occurs when charge carriers in the material are holes. Now, the transverse electric field E_y is from right to left, and exerts force F_E on the positive ions, and therefore on the bar, to the left. However, the Ampere force does not depend on the Hall coefficient, it is still to the right. How is this possible?

The answer is the secret of the holes. Why only holes can give rise to superconductivity. Figure 23.3 explains it. The key is the right panel.

The textbooks explaining the Hall effect show the left and center panels of Fig. 23.3. None shows the right panel. However, the center panel is a fiction. 'Holes' are not real particles, they are a theoretical construct that is useful to describe some phenomena, but can be completely misleading to understand other phenomena. Such is the case here.

The charges that move in a metal are always electrons of negative charge. Therefore, let us consider the right panel in the figure.

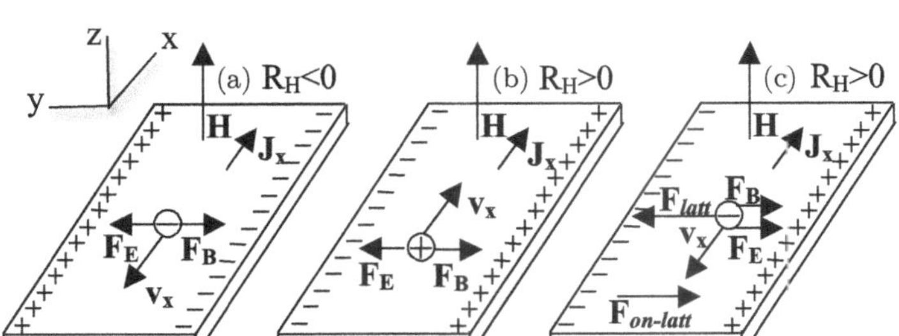

Fig. 23.3 Hall bar with negative Hall coefficient (left panel) and with positive Hall coefficient (center and right panels). For explanation, see text.

Instead of drawing holes moving in the $+\hat{x}$ direction, we draw electrons moving in the $-\hat{x}$ direction. We draw the electric and magnetic forces acting on electrons, F_E and F_B, both acting to the right. But the current, and hence the electrons, are moving along the x axis in the stationary state, there is no transverse current. This means that *there is another force in the transversal direction acting on electrons* so that they move in a straight line. That other force is the sum of electric and magnetic forces, F_E and F_B, acting in opposite direction, that is from right to left. We call it F_{latt} in the figure.

F_{latt} is a force that the lattice of ions exerts on electrons when they move in this configuration. F_{latt} exists when the electronic band is almost full, as the right panel of Fig. 16.2 shows, it doesn't exist when the band is almost empty, Fig. 16.2 left panel. When the band is almost full, we talk about holes rather than electrons, Fig. 16.3. But no matter what we talk about, the moving charges are always electrons.

We will explain in a moment how this difference arises. But before that, let us go back to Fig. 23.3 and remember Newton's third law. If the lattice of ions is exerting a force F_{latt} on the electrons, the electrons in turn exert an equal and opposite force on the ions. We call that force $F_{\text{on-latt}}$ in the figure, force on the lattice, and it acts from left to right.

The force $F_{\text{on-latt}}$, subtracting from it the force of the electric field F_E acting on the ions from right to left (Fig. 23.2 right panel) that is half in magnitude, gives a net force on the ions to the right that is precisely the Ampere force on the bar. This resolves the mystery of Fig. 23.2.

What is the relevance of this to superconductivity? Very simple. In words, when R_H is positive and there is conduction in electric and magnetic fields as the figure shows, there is a transfer of momentum between electrons and ions (in transverse direction) due to these forces F_{latt} and $F_{\text{on-latt}}$, that does not depend on the presence of impurities and irreversible scattering processes. It is a coherent interaction between the wavefunction of the electron and the periodic lattice of ions. This reversible momentum transfer between electrons and ions is necessary to explain how the momentum of the

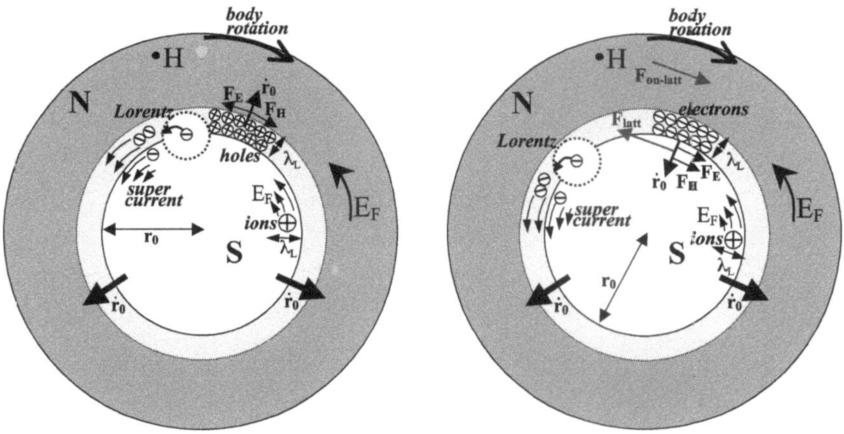

Fig. 23.4 Meissner effect again. The outflow of electrons that gives rise to the Meissner current is indicated with the dotted circle, that shows that when the orbit expands the electron acquires counterclockwise velocity due to the Lorentz force. The backflow of normal electrons is indicated as outflow of holes in the left panel and as inflow of electrons in the right panel. Both versions are equivalent, the balance of forces and how the body rotation is generated is explained in the text.

supercurrent and that of the body are compensated in the Meissner effect. Or, in the inverse process, when the superconductor goes normal, how the momentum of the supercurrent is transferred to the body without irreversible scattering processes that generate entropy.

Figure 23.4 shows how Fig. 23.3 applies to the transition to superconductivity in the presence of a magnetic field in the interior of the normal metal, i.e. the Meissner effect. The left panel of Fig. 23.4 corresponds to the central panel of Fig. 23.3 and the right panel of Fig. 23.4 corresponds to the right panel of Fig. 23.3.

The electric field in Fig. 23.4 is the Faraday field that is generated due to the outward motion of the phase boundary. On the upper part of the figure, it points left, like the electric field in the Hall bars of Fig. 23.3 with positive R_H. The current J_x in Fig. 23.3 is the backflow of normal charge in Fig. 23.4, which we can interpret either as an inflow of electrons (right panel of Fig. 23.4) or an outflow of holes (left panel of Fig. 23.4). In both interpretations, the azimuthal forces are balanced: for the holes directly, for the electrons including the force F_{latt}. The important point is that in either of the two interpretations the azimuthal forces have to be balanced in order for the flow to be exactly radial, with no azimuthal component. This occurs automatically because the Faraday field is given by Eq. (22.3), giving rise to an electric force

$$F_E = e\frac{\dot{r}_0}{c}H_c \tag{23.2}$$

where $\dot{r}_0$ is the radial velocity of the phase boundary between the phases, which equals the speed of backflow of the normal carriers, so the Lorentz force on them is

$$F_H = \frac{e}{c}\vec{\dot{r}}_0 \times \vec{H} \tag{23.3}$$

with $H = H_c$ the magnetic field at the phase boundary, hence both forces have equal magnitude.

The fact that there is no azimuthal component to this flow of normal charge guarantees that it is not necessary that there are irreversible scattering processes between charge carriers and impurities to transmit azimuthal momentum from electrons to the body. The body acquires rotational momentum in clockwise direction due to $F_{\text{on-latt}}$ on the right panel of Fig. 23.4, subtracting the direct force of the Faraday field on the ions that points to the left and is half in magnitude, giving a net force in clockwise direction that is the *Ampere force* of Fig. 23.2, that pushes the Hall bar to the right despite the fact that E_y points to the left, and makes the cylinder in Fig. 23.4 rotate clockwise, compensating the momentum that the electrons in the supercurrent are acquiring in counterclockwise direction.

As promised, Fig. 23.5 shows the transition in opposite direction, that is the transition from a superconductor that excludes magnetic field and has a surface current to the normal metal with no surface current and magnetic field in the interior. The processes are the same as in Fig. 23.4, in opposite direction.

The Meissner current stops without collisions that would generate entropy, it stops because electronic orbits are shrinking and in the process the Lorentz force pushes in opposite direction to the initial velocity. The momentum of electrons in the supercurrent is transmitted to the body through the same mechanism as in the inverse

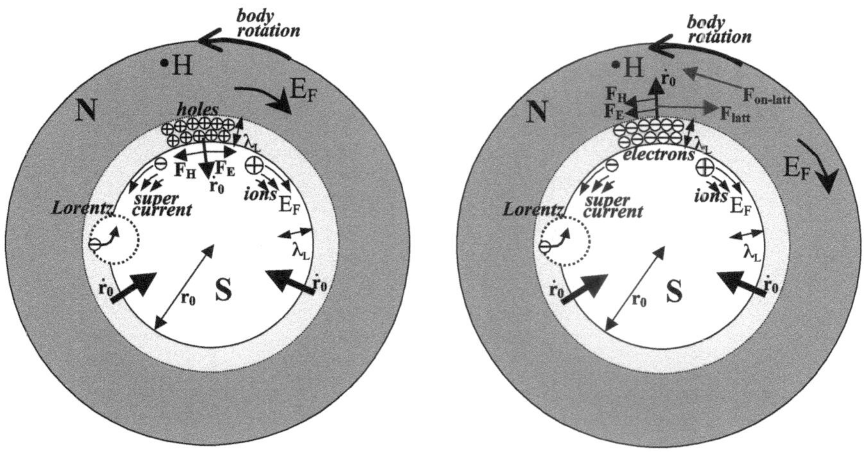

Fig. 23.5 Transition from superconductor to normal metal, that is, inverse Meissner effect. The Meissner current stops due to the Lorentz force acting on the shrinking dotted circles. The body acquires rotation in the same direction that the electrons were flowing in the supercurrent that stopped, that is in counterclockwise direction, due to the force $F_{\text{on-latt}}$. The two panels show the backflow of normal charge with holes and with electrons, as in Fig. 23.4 but in opposite direction.

process (Meissner effect), that is through the counterflow of normal electrons in a band that is almost full. This answers "The simplest question in superconductivity, that BCS doesn't answer" discussed in Chapter 6: how does the supercurrent stop in a superconductor that is becoming normal without dissipating heat and respecting momentum conservation?

It remains to be understood: why do holes behave the way they do? Or equivalently why do electrons near the top of the band behave differently than electrons near the bottom of the band? Let us consider Fig. 23.6.

The upper left panel shows the energy of electrons as function of the variable k. The quantity $\hbar k$, with $\hbar$ Planck's constant, is what is called the 'crystal momentum' of the electron. It is similar, but not the same, as the mechanical momentum of the electron. The upper panel is the same as in the bands shown in Fig. 16.2, where we said that the crystal momentum is inversely proportional to the

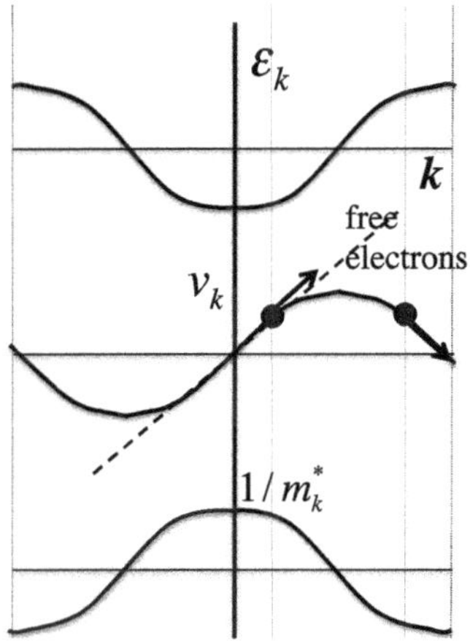

$$\frac{d}{dt}(\hbar\vec{k}) = \vec{F}_{ext} = e\vec{E} + \frac{e}{c}\vec{v}_k \times \vec{H}$$

$$v_k = \frac{1}{\hbar}\frac{\partial \varepsilon_k}{\partial k}$$

$$\vec{a}_k = \frac{d}{dt}\vec{v}_k = \frac{1}{\hbar^2}\frac{\partial^2 \varepsilon_k}{\partial k^2}\frac{d}{dt}(\hbar\vec{k}) = \frac{1}{\hbar^2}\frac{\partial^2 \varepsilon_k}{\partial k^2}\vec{F}_{ext} \equiv \frac{1}{m_k^*}\vec{F}_{ext}$$

$$\vec{a}_k = \frac{1}{m_e}\vec{F}_{total} = \frac{1}{m_e}(\vec{F}_{ext} + \vec{F}_{latt}) = \frac{1}{m_k^*}\vec{F}_{ext}$$

$$\vec{F}_{latt} = [\frac{m_e}{m_k^*} - 1]\vec{F}_{ext}$$

$$\frac{1}{m_k^*} = \frac{1}{\hbar^2}\frac{\partial^2 \varepsilon_k}{\partial k^2}$$

Fig. 23.6 Left panels, upper, center, lower: energy, velocity and effective mass of an electron in state k as function of the crystal momentum k. To the right, the equations that determine the change in the crystal momentum and electronic velocity with time when an external force $\vec{F}_{\text{ext}}$ acts. $\vec{F}_{\text{latt}}$ is the force that the lattice exerts on the electron and $\vec{F}_{\text{total}}$ is the total force exerted on the electron.

wavelength of the electron. When k is small, the wavelength is much larger than the distance between ions and the electron is not strongly affected by the ionic potential, it behaves more or less like a free electron. Instead, when k increases and we approach the top of the band, the wavelength becomes smaller until it becomes comparable to the distance between ions, and in that regime the ionic potential strongly affects the wavefunction of the electron.

The first equation in the figure is a basic formula of Bloch's theory of solids: when an external force F_{ext} is applied, for example, electric and magnetic fields, the crystal momentum $\hbar\vec{k}$ changes with time as if it was the mechanical momentum of the electron: its time derivative equals the external force. But it is more subtle, because the crystal momentum is not the same as the mechanical momentum.

The mechanical momentum of the electron is its mass m_e multiplied by its velocity v_k, and that velocity is given by the slope of the curve ϵ_k as function of k, as the first equation in the blue box in the figure and the graph in the central panel shows. As function of positive k, the figure shows that v_k first increases and then it decreases.

When an external force is applied, k changes according to the first equation. Let us assume the external force is such that $d(\hbar k)/dt$ is a positive constant, and the electron is initially in the state $k = 0$. As time progresses, k increases and its velocity v_k increases, as the central panel shows. But then v_k reaches a maximum and it starts to decrease, always subject to the same external force. The reason is, another force starts to act, an internal force due to the potential of the ions, in direction opposite to the external force.

What this tells us is that initially the electron behaves as a free particle, when the external force is applied the velocity increases in the direction of the applied force. The acceleration a_k is given by

$$a_k = \frac{d}{dt}v_k = \frac{1}{\hbar}\frac{d}{dk}v_k\frac{d}{dt}(\hbar k) = \frac{1}{\hbar}\frac{d}{dk}(v_k)F_{\text{ext}} = \frac{1}{m_k^*}F_{\text{ext}} \qquad (23.4)$$

so that it is proportional to the slope of the curve v_k as function of k. This slope, the inverse of the 'effective mass' of the electron, m_k^*, is positive for $k = 0$ but as k increases it decreases and then it turns negative, as the third panel shows.

This means that the acceleration, which is given by the external force divided by the effective mass, is parallel to the external force when k is small and antiparallel when k is large. The constant of proportionality between acceleration and external force, $1/m_k^*$, is the curvature of the function ϵ_k versus k. For a free electron, $m_k^* = m_e$, the free electron mass, and for electrons near the bottom of the band $m_k^* \sim m_e$. Instead, for electrons near the top of the band m_k^* is negative.

As we said, the acceleration (and m_k^*) change sign because there is another force acting in addition to the external force, an internal force that we call F_{latt}. As the figure shows, we can express F_{latt} in terms of m_e, m_k^* and the external force:

$$\vec{F}_{\text{latt}} = \left[\frac{m_e}{m_k^*} - 1\right]\vec{F}_{\text{ext}}. \qquad (23.5)$$

When $m_k^* = m_e$, $\vec{F}_{\text{latt}} = 0$. When $m_k^* < 0$, $\vec{F}_{\text{latt}}$ is maximum, and in direction opposite to the applied force $\vec{F}_{\text{ext}}$.

We will not go into more details. It can be shown mathematically [4] that when $m_k^* > 0$ for electrons at the Fermi level (the maximum energy of occupied states) the total internal force is zero, as the left panel of Fig. 23.3 shows. When the Fermi level is above the middle of the band m_k^* turns negative and the total F_{latt} is not zero (Fig. 23.3 right panel). In that regime, there is transfer of momentum between the charge carriers and the ions, a necessary condition to satisfy momentum conservation in the Meissner effect and in the inverse transition.

Since we are at it, let us understand why the Hall coefficient is negative when the Fermi level is close to the bottom of the band and positive when it is close to the top of the band. The books explain it either with words and/or figures using holes that don't explain anything, or with very complicated mathematical explanations. Figure 23.7 explains it in a simple way.

Figure 23.7 shows the electrons in the Hall bars of Figs. 23.2 or 23.3. The Lorentz force that the magnetic field exerts on these electrons is

$$\vec{F}_{\text{ext}}^{H} = e\frac{\vec{v}_k}{c} \times \vec{H}. \tag{23.6}$$

The direction of this force depends on the sign of the velocity, $\vec{v}_k$, so it has opposite sign for k positive and negative. In addition, the acceleration a_k^H that this force produces depends on the sign of the effective mass m_k^*:

$$a_k^H = \frac{1}{m_k^*}\vec{F}_{\text{ext}}^{H} \tag{23.7}$$

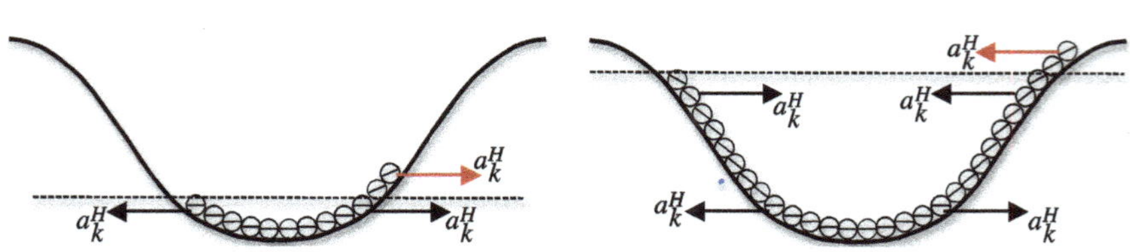

Fig. 23.7 Explaining the opposite sign of the Hall effect when the band is almost empty and almost full. The red arrows indicate the direction of the net flux of electrons in the transverse direction. In the Hall bars of Figs. 23.2 and 23.3, this flux gets cancelled when charge accumulates on the side edges of the bar and the transverse electric field is established.

Because a current is flowing, the occupation is not symmetric in k: there are more electrons on the positive k side than on the negative. The figure shows the direction of the transverse acceleration (y direction in Figs. 23.2 and 23.3) depending on the position of the electron in the band. In the region where the occupation is symmetric in k and $-k$, the accelerations are in opposite directions and there is no net flux. Therefore, the direction of flux in the transverse direction is determined by the electrons that are above the dashed lines, that denote the position of the Fermi level. The red arrows indicate the direction of net flux, which is opposite when the band is almost empty and almost full because the sign of the effective mass m_k^* for electrons close to the Fermi level is opposite in both cases, as we saw in the lower left panel of Fig. 23.5.

Returning to Fig. 23.1 that lists the differences in the behavior of an electron close to the bottom of the band (bonding) and an electron close to the top of the band (antibonding), difference number 9 says that the first is "Detached from lattice" and the second "Transfers momentum to lattice". As we saw in this chapter, in the metal–superconductor or superconductor–metal transition in the presence of a magnetic field, it is necessary to transfer momentum between electrons and the lattice of ions, and this transfer has to occur in a reversible way. This occurs when the charge carriers in the normal state at the Fermi energy have negative effective mass, and doesn't occur when they have positive effective mass. This explains the correlation found by Kikoin and Lasarew already in 1932 and rediscovered by various other researchers later, that positive Hall coefficient favors superconductivity.

Finally, let us return to the discussion of Chapter 10 on the relation between the Meissner effect and Alfven's theorem, the statement that in a perfectly conducting fluid magnetic field lines are frozen into the fluid and move together with the fluid. In Fig. 23.8, we reproduce Fig. 10.3, but now on the left panel we attached a mirror image of the jet of fluid, flowing to the left. Now on the left panel the magnetic field lines bend in both directions, as in the Meissner effect on the right panel. The white arrows on the right panel show the outflow of a perfectly conducting fluid composed of electrons and holes, as

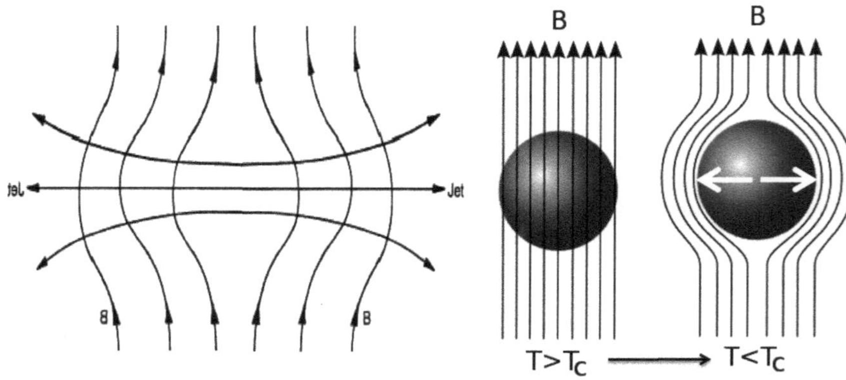

Fig. 23.8 Left panel: the right half is from a picture in Ref. [1] of Chapter 10, with caption: '*An example of Alfven's theorem. Flow through a magnetic field causes the field lines to bow out.*' I copied it, flipped the copy horizontally, and juxtaposed it to the left. On the right panel, a picture of the Meissner effect: as the temperature is lowered, it '*causes the field lines to bow out*'. The white arrows show the predicted 'Jets'.

shown on the left panel of Fig. 23.4, carrying the magnetic field lines out. Looking at this figure, one may ask: why haven't plasma physicists explained to solid state physicists how the Meissner effect works already long ago?

The answer is, the situation shown on the left panel of Fig. 23.8 cannot occur in real plasmas. If it is a charge neutral plasma, the jet does not carry charge but it carries mass, and the situation on the left panel of Fig. 23.8 would leave the interior devoid of mass, which can't happen.

The reason this can happen in a solid is because holes do not carry real positive mass, they carry negative mass. As illustrated in Fig. 23.9, holes flowing to the right is equivalent to electrons flowing to the left. This was also shown in a different way in Fig. 5.1. It is the electrons that carry the real mass. Hence, when both electrons and holes flow out, as shown in Fig. 23.4, there is no net outflow of mass. What does flow out is 'effective mass' [5], leading to the conclusion that when a metal goes superconducting the carriers lower their effective mass, which is in agreement with experimental observations in superconductors [6].

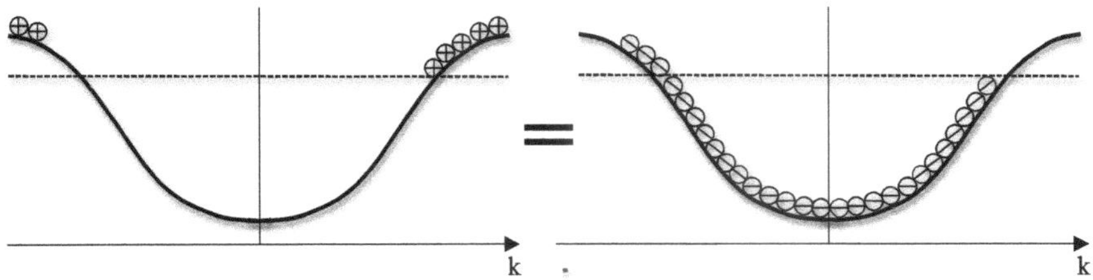

Fig. 23.9 Holes flowing in the positive k direction (left panel) corresponds to electrons flowing in the negative k direction (right panel).

The transport of electricity from a normal conducting wire into a superconducting one also illustrates vividly that Alfven's theorem is at play. Figure 23.10 shows the current streamlines and the magnetic field lines in such a situation. resulting from solution of the London equations. It clearly shows that a perfectly conducting fluid is carrying the field lines with it, as required by Alfven's theorem. It is remarkable that the behavior shown in Fig. 23.10 has been known for over 80 years. yet within the generally accepted theory of superconductivity Alfven's theorem plays no role in it.

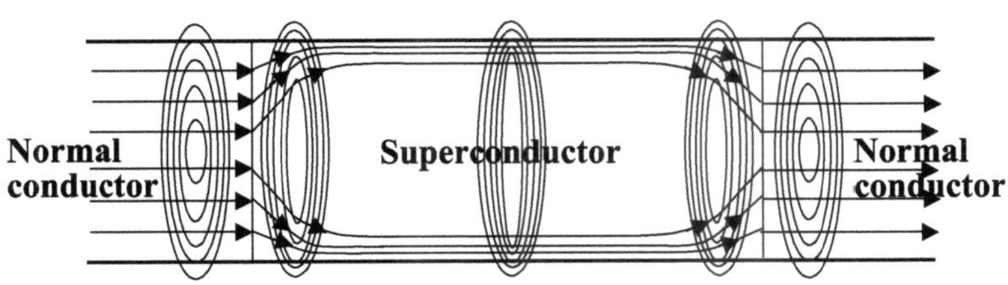

Fig. 23.10 Current distribution and magnetic field lines (vertical circles) in a superconducting cylindrical wire fed by normal conducting leads.

The analysis of this chapter leads to a conclusion that I believe is absolutely general: *if there are no charge carriers near the Fermi energy that have negative effective mass in a material, that material cannot be a superconductor at any temperature.* There are charge carriers of negative effective mass near the Fermi level when the band is almost full, in which case the charge carriers are holes.

References

[1] J. E. Hirsch, Why holes are not like electrons. II. The role of the electron–ion interaction, *Phys. Rev. B* **71**, 104522 (2005).
[2] J. E. Hirsch, The disappearing momentum of the supercurrent in the superconductor to normal phase transformation, *Europhys. Lett.* **114**, 57001 (2016).
[3] J. E. Hirsch, On the reversibitity of the Meissner effect and the angular momentum puzzle, *Annals of Physics* **373**, 230 (2016).
[4] J. E. Hirsch, Momentum of superconducting electrons and the explanation of the Meissner effect, *Phys. Rev. B* **95**, 014503 (2017).
[5] J. E. Hirsch, How Alfven's theorem explains the Meissner effect, arXiv:1909.11443 (2019).
[6] M. V. Klein and G. Blumberg, Effective mass and color change, *Science* **283**, 42 (1999).

Chapter 24

BCS versus Hubbard versus Holes

We have talked in this book about three different theoretical frameworks to understand superconductivity in nature: (1) BCS theory, (2) theories based on the Hubbard model, and (3) Hole superconductivity. Which is right? Which is wrong?

First possibility: it is certainly possible that all three could be incorrect, that none of these describes any real material.

Second: it is possible that BCS describes some materials. If so, it is also possible that the Hubbard model describes other materials. In that case, it is not possible that the theory of hole superconductivity applies to any material.

Third, it is possible that the Hubbard model describes some materials while neither BCS nor hole superconductivity do.

Fourth, it is possible that the theory of hole superconductivity applies to all materials. In that case, BCS theory is not applicable nor are Hubbard models.

What is not possible is that the theory of hole superconductivity describes some materials and not others. If it describes some, it describes all. If not, it doesn't describe any.

This means that the theory of hole superconductivity is falsifiable by finding a single material that clearly cannot be explained by that theory. In contrast, neither BCS theory nor Hubbard model approaches can be proven wrong by finding materials to which they don't apply. One can always say that they are a different class of 'unconventional' materials.

When I discuss with colleagues and argue that only the hole theory can describe superconductivity, they tell me: how come, don't you agree that the attractive Hubbard model describes superconductivity, hence could describe the superconductivity of some materials, even if only materials that have not yet been discovered?

The attractive Hubbard model is simply a model of electrons where one assumes that there is an attractive rather than repulsive interaction when two electrons occupy the same orbital. It is a totally unphysical assumption, not representing what occurs in nature. But let's assume that is not so, that the model is valid to describe some material. Then, it is easy to show mathematically that electrons pair to lower their (potential) energy, and the ground state of this model has a wavefunction of the same form as given by BCS, Eq. (8.1). Such a system has phase coherence, an energy gap between the ground state and excitations, a density of states of the form that is measured experimentally by tunneling such as Fig. 14.2, etc. It certainly describes a superconductor, doesn't it?

I say no. I say that a system described by the attractive Hubbard model cannot exhibit the Meissner effect, simply because it does not expel charge. It does not have the means to expel a magnetic field, and it certainly cannot do it conserving momentum. So, if we cool a metal described by the attractive Hubbard model in the presence of a magnetic field, the field would remain inside and the system would never reach the superconducting state, it would remain in the metastable normal state forever. That certainly does not describe a real superconductor in nature, all superconductors expel magnetic fields.

On the other hand, over the past 30 years, there have been a great variety of theories based on the *repulsive* Hubbard model to describe superconductivity in the cuprates as well as in other systems that supposedly have so called "strongly correlated electrons" and are classified as 'unconventional superconductors'. As we mentioned earlier, this was propelled to a large extent by a very influential paper by Phil Anderson in 1987 [1], titled "The Resonating Valence Bond State in La_2CuO_4 and Superconductivity", where he proposed that electrons in cuprates are 'spin liquids' described by the Hubbard model, that have preformed 'singlet pairs' in the normal state that condense and give rise to the superconducting state. That article by Anderson has 6002 citations, and articles citing it have been cited 250,306 times. The theory based on that paper by Anderson is called 'RVB', initials of 'Resonating valence bond' state.

At that time, 1987, I found that paper interesting because only two years earlier I had published a paper [2] proposing that the Hubbard model at low temperatures describes a superconducting state induced by magnetic singlet pairs, and I had presented numerical calculations that appeared to support that possibility. I didn't propose it in relation to the cuprates, that had not yet been discovered in 1985, but thinking about other (low T_c) superconductors called 'heavy fermion systems' that showed properties that appeared to suggest an anisotropic pair wavefunction (as was shown in Fig. 3.2), that would make the material 'unconventional'. In his 1987 paper, Anderson cites my 1985 paper saying "*Hirsch has shown numerically that the simple square lattice probably retains a magnetization but that finite U may favor the RVB state*", and "*Hirsch's simulations in these references are very suggestive confirmation of the RVB model of superconductivity.*"

The 1987 Anderson paper motivated a large part of the scientific community to concentrate in studying the new supposed state of matter, 'spin liquid', and the possible resulting superconductivity in the Hubbard model in the ensuing 30+ years, to the present. The fact that shortly after Anderson's paper was published it was determined experimentally that in the normal state the cuprates have antiferromagnetic order [3] and therefore *are not* a 'spin liquid' made no dent. Physicists simply assumed that cuprates are "almost" 'spin liquids', no big deal, the same Anderson physics applies.

Besides, from another angle, another large part of the scientific community focused also on the Hubbard model but with a more traditional viewpoint, assuming that the normal state is not an exotic 'spin liquid' but rather a garden variety 'Fermi liquid', the normal state of an ordinary normal metal, and that superconductivity results from 'spin fluctuations' induced by the Hubbard repulsion U that create singlet pairs in this system. Scalapino, mentioned in Chapter 2, was and is one of the principal proponents of this point of view. In the first paper where this was proposed in 1986, I was a coauthor [4], once again it was before the cuprates had been discovered, focused on metals with 'heavy fermions'.

Without intending to boast I would like to point out that at that time I was one of the most experts physicists on the Hubbard model. In a book published in 1992 [5] with reprints of selected articles on the Hubbard model, four of my papers are included, more than of any other scientist, three of Scalapino and two of Anderson. I mention this so that it is clear to the reader that my current disinterest in the Hubbard model is not because I am not familiar with it. By 1988, I had published about 20 papers on that model. On the other hand, Scalapino had published six articles on the Hubbard model by 1988 (five in collaboration with me) and went on to publish 75 more from 1989 to the present.

In late 1988, I came to the conclusion that neither my 1985 proposal [2] nor Scalapino and coauthors' proposal of 1986 [4] nor Anderson's of 1987 [1] nor the Hubbard model have anything to do with superconductivity of either the cuprates or any other material. Instead, I would estimate that more than 90% of physicists in the United States continue to be convinced that the Hubbard model describes superconductivity in the cuprates, as well as in other superconductors considered to be 'unconventional' [6]. Besides Anderson's and Scalapino's approaches, various other approaches based on the Hubbard model have been proposed: *d*-density wave, loop currents, quadrupolar fluctuations, broken time-reversal symmetry, SO(5), quantum critical points, holographic superconductivity, electronic liquid crystals, triplet superconductors, $s_{+/-}$, etc, etc.

After 30 years, there is still no convincing evidence that the 'spin liquid state' exists in the Hubbard model or in nature, nor is there agreement on whether the repulsive Hubbard model has a superconducting ground state with paired electrons, or not.

Not to mention explaining the Meissner effect, or rotating superconductors, with the Hubbard model. Nobody has attempted it. The lack of success doesn't dissuade the practitioners of these activities. On the contrary, they are excited to learn that a model in appearance as simple as the Hubbard model can be so complicated that it has not been understood after so many years and effort.

Since 1988 I say that the Hubbard model is not physical because it does not take into account charge asymmetry, which is an essential part of physical reality. It has electron-hole symmetry, that makes it an artificial model not suited to describe natural phenomena.

In the past 30 years, 5,000 articles with the words "Hubbard model" have been published, that have been cited 94,000 times. 18,000 articles on the subject "Hubbard model" that have been cited 415,000 times. None of those articles contributes anything to the understanding of superconductivity in cuprates nor in any other material in my opinion.

Among researchers in superconductivity the conviction that magnetism is intimately related and somehow causes the superconductivity of the cuprates and other 'unconventional' superconductors is practically universal. At the same time, it is universally believed that this has absolutely nothing to do with the superconductivity of elements and simple compounds, where superconductivity supposedly originates in the BCS electron-phonon interaction. Nobody has attempted to explain why nature has chosen not to use 'spin fluctuations' or magnetism to generate superconductivity in at least some elements or simple compounds. When 'spin fluctuations' are mentioned in relation with the elements is to explain why some element does not exhibit superconductivity. In my view, this defies the most elementary logic.

If BCS believers are asked, how many different 'unconventional' mechanisms of superconductivity are there, what is the answer? The most common one is, we don't know. Nobody will say "none", everybody agrees that there are materials that cannot be described by BCS and electron–phonon interaction [7]. Also practically nobody will say that there is just one unconventional mechanism besides the BCS conventional one. Because once the possibility of more than one mechanism is allowed. since there are a variety of classes of materials called unconventional that are very different among themselves, different unconventional mechanisms are explored to describe each of them. The general belief is that there are many unconventional mechanisms of superconductivity besides the conventional one, BCS–electron–phonon.

The concept itself seems absurd. If there really are several different mechanisms of superconductivity, why would the BCS one deserve a special role? What makes BCS with the electron–phonon interaction so special, other than the historical accident that it was the first mechanism proposed? And along the same lines, why think that all elements are described by BCS and none is described by one of the unconventional mechanisms?

Science generally progresses in the direction of unification and simplicity. In the science of superconductivity the path has been exactly opposite. From one single mechanism of superconductivity that existed between 1957 and ∼1980 and was believed to be universal, now it is believed that there are several mechanisms and different explanations for the great variety of superconducting materials known now.

Going back to BCS, I say that BCS theory with the electron–phonon interaction pairing the electrons, or with any other mechanism pairing the electrons, for example, a magical attractive Hubbard interaction, does not have a mechanism to expel magnetic fields that satisfies the law of momentum conservation. That to expel a magnetic field it is indispensable that there is flow and backflow of charge in the radial direction, that is in the direction in which the magnetic field is expelled. That the motion of magnetic field lines is necessarily associated with the motion of electric charge in the same direction. As Alfven's theorem taught us. Neither BCS nor Hubbard models say that.

How can it be proven that I am wrong? For example, finding another different mechanism to expel magnetic fields. Nobody has done it, I am convinced it is impossible. BCS defenders say that BCS surely can do it, even if nobody has explained how. That BCS describes a system where the lowest energy state has the magnetic field excluded, therefore physical systems will find a way to reach the state of lowest energy. They always do that. It is not necessary to know exactly how they do it.

That is nonsense. It is not true that physical systems spontaneously evolve to states of lower energy. They do it if they have a path to do it, guided by a dynamics that satisfies the laws of physics. If not, they don't do it.

It is obvious, but incomprehensible to those that believe in BCS. If all physical systems would evolve to states of lower energy, we wouldn't exist. There would not be mountains, everything would be plane. There would not be weather changes, earthquakes, tornados, or supernovas. Different elements would not exist, all atomic nuclei would have the number of neutrons and protons that give maximum stability, 28 protons and 34 neutrons, forming Nickel-62.

Those that believe in BCS will say those are classical examples, but BCS is governed by quantum mechanics, that is more complicated and less intuitive. I say, superconductors are macroscopic systems and Bohr's correspondence principle tells us that there has to be a continuous connection between the microscopic and macroscopic worlds. And that at the macroscopic level BCS theory can only expel magnetic fields if it violates momentum conservation and violates Faraday's law and violates conservation of entropy in reversible transformations. That is, it contradicts Newton's mechanics, Maxwell's electromagnetism, and the laws of thermodynamics.

BCS defenders say the theory has existed for more than 60 years, it describes many properties of superconductors, and it is impossible that it's wrong.

I say, Ptolemy's astronomical theory was considered valid for 1500 years, it explained many astronomical observations, and then it was discovered that it was totally wrong.

Finally, why am I convinced that the concepts associated with the theory of hole superconductivity discussed in this book describe physical reality?

There are simply too many coincidences, that I have been finding over many years, unexpected discoveries that convince me that this is so. Different parts of the theory are intimately related in a coherent way, in ways that were not at all obvious initially.

Summarizing, novel elements of this theory that *are not contained in other theories* are:

(1) The theory predicts that superconductivity can only occur when there are almost full bands, that is charge carriers that are holes. This agrees with observations in multitude of systems.
(2) It predicts that superconductivity is associated with lowering of quantum kinetic energy, coincident with what is known occurs in superfluid helium, another macroscopic quantum system.
(3) It predicts that metals expel negative charge when they become superconducting. This gives a natural mechanism for expulsion of magnetic fields, i.e. the Meissner effect.
(4) It predicts that superconductors are 'giant atoms' literally, as Fig. 17.1 shows, with more negative charge near the surface and more positive charge in the interior, just as microscopic atoms.
(5) It predicts that electronic orbits expand when the system becomes superconducting. This explains naturally the observed behavior of rotating superconductors.
(6) It predicts that there is macroscopic zero point motion in superconductors, with electronic orbits with angular momentum quantified with magnitude $\hbar/2$, the same as the intrinsic angular momentum of the electron.
(7) It explains the dynamics of how magnetic fields are expelled when metals become superconducting, and how the supercurrent stops without dissipation when a superconductor becomes normal.
(8) It explains how momentum is conserved in the metal-superconductor transition and vice versa, which depends crucially on the condition that the charge carriers are holes.

All of this follows from taking into account one property of the microscopic world that is undeniably real and universal: that when two electrons occupy an atomic orbital, they don't occupy the same orbital as a single electron. The orbital expands due to

Coulomb repulsion, the quantum kinetic energy decreases due to the orbital expansion, and negative charge is expelled radially outward due to the orbital expansion. This simple fact, that neither the Hubbard model nor BCS take into account, leads to the consequences (1)–(8) listed above. In 2001, I started to call models that describe this physics "Dynamic Hubbard models" [8], because one can think of them as Hubbard models where the value of U changes with the expansion of the orbital, and because I hoped that giving them that name would make them more acceptable to the physics community enthralled with the Hubbard model. It didn't work, so far.

The development of these ideas took me 30 years. It was not until the first 15 years had passed that for the first time I started to realize that this physics is intimately tied to the Meissner effect, the most fundamental property of superconductors. And that without this physics the Meissner effect cannot be explained. Before then, like all other physicists, I had not asked myself whether BCS theory explains the Meissner effect, I took it for granted.

These considerations, and the fact that the theory of hole superconductivity applies to all superconductors, not to some classes yes and to other classes no like BCS and Hubbard, convince me that the theory reflects the real world, while BCS and Hubbard theories reflect the world of the imagination of their believers rather than the real world.

Why is it important to decide whether BCS is correct, or Hubbard is correct, or Holes are correct? One of the main reasons is discussed in the following chapter.

References

[1] P. W. Anderson, *Science* **235**, 1196 (1987).
[2] J. E. Hirsch, Attractive interaction and pairing in Fermion systems with strong on-site repulsion, *Phys. Rev. Lett.* **54**, 1317 (1985).
[3] D. Vaknin *et al.*, Antiferromagnetism in La_2CuO_{4-y}, *Phys. Rev. Lett.* **58**, 2802 (1987).

[4] D. J. Scalapino, E. Loh, Jr., and J. E. Hirsch, d-wave pairing near a spin-density-wave instability, *Phys. Rev. B* **34**, 8190(R), 1986.
[5] *The Hubbard Model — A Reprint Volume*, edited by Arianna Montorsi, World Scientific, Singapore, 1992.
[6] P. A. Lee, N. Nagaosa, and X. G. Wen, Doping a Mott insulator: Physics of high-temperature superconductivity, *Rev. Mod. Phys.* **78**, 17 (2006).
[7] M. L. Cohen, *Mod. Phys. Lett. B* **24**, 2755 (2010).
[8] J. E. Hirsch, Dynamic Hubbard model, *Phys. Rev. Lett.* **87**, 206402 (2001).

Chapter 25

How to find and not find high temperature superconductors

If superconductivity was properly understood, and that understanding was generally shared, it would be much easier to find new materials that superconduct at higher temperature, ideally at room temperature, since the search for new superconducting materials would be guided by that understanding. This would enormously broaden the practical applications of superconductors in technology and in everyday life. Imagine how backward the electronic industry would be today if semiconductors only would work at liquid nitrogen temperature, 77 K. How would we keep our cellphones that cold?

Since time immemorial physicists have tried to develop theoretical criteria to guide the search for new superconductors at higher temperatures, particularly after BCS. None have worked. All the discoveries of new superconductors were either by accident or following empirical criteria like Matthias', without a theoretical basis.

Matthias said in his 1971 article titled "*The search for high-temperature superconductors*" [1]: "*the theories of these last 21 years should have also been able to show a way (if one exists) to increase the critical temperature. But until this day there has not been a shred of evidence for this expectation. I can think of no other field in modern physics in which so much has been predicted without producing a single experimental success.*"

The same continues to be true today, 48 years later. During these 48 years, the maximum critical temperature increased by a factor of 7, *without any help from theories.*

With 'unconventional' mechanisms based on the Hubbard model, there have not been many attempts at predicting new materials. The few that have been made, for example that iridium compounds would be high temperature superconductors, have certainly not been successful.

With the conventional BCS mechanism, one prediction that has received a lot of attention, due originally to Ashcroft, is that hydrogen-rich compounds should be high temperature superconductors. Recently published papers claim [2, 3] that hydrogen compounds at very high pressure show superconductivity close to room temperature. Those results have not been confirmed, I believe they are not correct. There continue to be multitude of papers predicting superconductivity in a variety of compounds based on BCS theory. Sometimes it is difficult to understand from the article if it is talking about experiments or about numerical calculations, when they say that a compound shows superconductivity. In the mind of many theorists of today, if a numerical calculation says that a compound is a superconductor, the compound is a superconductor. It is not necessary to verify it experimentally. And if the experiment is performed and the calculation is not verified, the material is making a mistake.

I say, that is not science.

The theory of hole superconductivity gives well defined criteria for how to find high temperature superconductors, namely:

(1) Charge carriers have to be holes. If there are no holes, there is no superconductivity.
(2) Conduction through anions, i.e. negative ions from the right side of the periodic table. The more negative the better.
(3) Much negative charge in the region where holes conduct and give rise to superconductivity.
(4) The mass of the ions is unimportant.
(5) Structures with planes are favorable because they allow to accumulate substantial negative charge in the planes, compensating it with positive charge outside the planes.

Those are general criteria that should be useful in the search, and in saving wasted efforts into compounds that have no chance of being superconductors at high temperatures, because they have no holes, or no negative ions, etc.

Let us consider a criterion that was proposed many years ago, before BCS, to understand why some materials are superconductors and others are not, the Meissner–Schubert criterion (the same Meissner that discovered the Meissner effect).

Meissner and Schubert in 1943 calculated [4] the volume occupied per conduction electron in the elements, and found that superconductors have volume per conduction electron that is generally smaller than that of non-superconductors, as Fig. 25.1 shows.

It is interesting that in a paper by the experimental physicist Eduard Justi in 1946 [5], he comments about this Meissner–Schubert criterion, saying: "*this rule has an obvious truth content and merits careful attention*". In the same article, Justi considers the possible

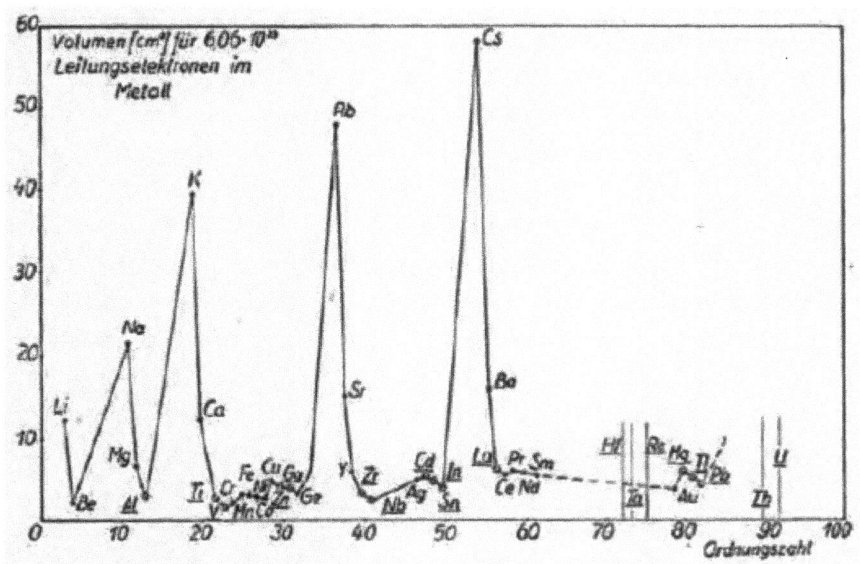

Abb. 2. Supraleiter (unterstrichen) und Volumen für Leitungselektronen.

Fig. 25.1 Meissner–Schubert diagram. The vertical axis is the volume occupied per conduction electron and the horizontal axis is the atomic number of the element.

correlation between the Debye temperature θ_D of ionic vibrations (inversely proportional to the square root of the ionic mass) and superconductivity, and says "*Considering the values of* θ_D *as criterion to distinguish between superconducting and non-superconducting metals, we find that there is no distinction of the superconductors with respect to the values of* θ_D*, on the contrary those values cover the entire range of observed* θ_D *values from* 69 K *to* 400 K*, nor is there any relation between* θ_D *and the critical temperature* T_c *of the superconducting metals*". This is of course surprising within BCS theory (that didn't exist when Justi wrote his article) within which the formula for T_c, Eq. (8.2a), is proportional to the Debye temperature.

In 1997, I did a statistical study of the possible relation between different normal state properties of the elements and superconductivity [6], assuming we know nothing about the mechanism that causes superconductivity. I considered 13 properties, not including the volume per electron considered by Meissner–Schubert because it didn't occur to me to do so and because I wasn't aware of their paper at the time. I found what Fig. 25.2 shows.

The Kolmogoroff–Smirnoff index that I calculated (vertical scale in the figure) measures the statistical dependence between the two variables considered, that is (1) existence or nonexistence of superconductivity in a given element, and (2) value of a given normal state property. The results of Fig. 25.2 show that the Hall coefficient is by a large margin the normal state property that is most closely associated with the existence or non-existence of superconductivity among all the properties considered. More specifically (not shown in Fig. 25.2), that positive Hall coefficient favors superconductivity and negative Hall coefficient disfavors superconductivity. In contrast, the quantities that BCS predicts should be associated with superconductivity, such as Debye temperature, electronic specific heat and electrical conductivity, don't have a significant association with superconductivity according to Fig. 25.2.

As we saw in earlier chapters, positive Hall coefficient means holes in almost full bands, which implies a lot of electrons in the band, which indicates that the available volume for each electron is small.

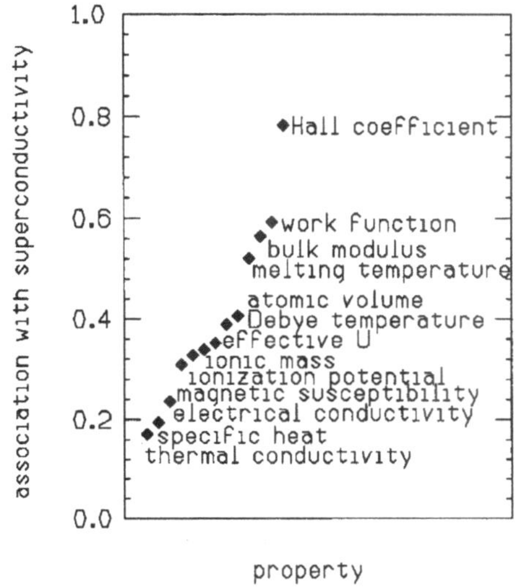

Fig. 25.2 Degree of association of different normal state properties of the elements with existence of superconductivity, measured by the Kolmogoroff–Smirnoff index [6]. From 0 to 1, it means complete independence or complete dependence. The horizontal axis is arbitrary.

Note that this is exactly what Meissner and Schubert independently found 50 years earlier!

Nowadays a methodology to find new superconducting materials called "machine learning" is becoming very popular [7]. Copying from Wikipedia, "*Machine learning (ML) is the scientific study of algorithms and statistical models that computer systems use to perform a specific task without using explicit instructions, relying on patterns and inference instead … Machine learning algorithms build a mathematical model based on sample data, known as "training data", in order to make predictions or decisions without being explicitly programmed to perform the task.*" In a way, it is what I did in that 1997 study [6]. The computer is given the information on all those normal state properties of the elements, it is told which are superconductors and which are not, and it is asked to find which properties of the normal state are more indicative of superconductivity, to guide future searches of new superconductors.

Personally I think that this procedure is not very useful. It is somehow to throw in the towel. What is useful is to understand which properties favor superconductivity and why. In this book, we have seen why in Fig. 25.2 the Hall coefficient is head and shoulders above all the other properties.

A lot remains to be done in refining the criteria (1)–(5) that I gave above to find new superconductors at high temperatures. To make those criteria more specific, it is necessary to perform band structure calculations computing in detail certain electronic properties, taking into account those criteria. I believe that is feasible and should be successful, but it has never been done before. It is necessary to devote resources in that direction, which will not happen until the scientific community decides to call BCS into question.

Ten years ago I published an article titled "*BCS theory of superconductivity: it is time to question its validity*" [8]. I thought the time had come back then, but no. I hope the time has come now, and that this book contributes for it to be so. I invite the reader interested to learn more about the technical issues discussed in this book to look at the papers I published on the subject during the last 30 years, listed in the website https://jorge.physics.ucsd.edu/hole.html.

References

[1] B. T. Matthias, *Physics Today* **24**(8), 23 (1971).
[2] M. Somayazulu *et al.*, Evidence for superconductivity above 260 K in lanthanum superhydride at megabar pressures, *Phys. Rev. Lett.* **122**, 027001 (2019).
[3] A. P. A. P. Drozdov *et al.*, Superconductivity at 250 K in lanthanum hydride under high pressures, *Nature* **569**, 528–531 (2019).
[4] W. Meissner and G. Schubert, Sitz, Ber. Bayr. Akad. d. Wiss., *Nat. Abt.* **295** (1943).
[5] E. Justi, Elektrische Supraleifäihigkeit, *Naturwissenschaften* **33**, 292 (1946).
[6] J. E. Hirsch, Correlations between normal-state properties and superconductivity, *Phys. Rev. B* **55**, 9007 (1997).
[7] V. Stanev *et al.*, Machine learning modeling of superconducting critical temperature, *npj Computational Materials* **4**(29) (2018).
[8] J. E. Hirsch, BCS theory of superconductivity: It is time to question its validity, *Phys. Scripta* **80**, 035702 (2009).

Postface

In this book, I have exposed a new way to understand the phenomenon of superconductivity, that differs in many fundamental aspects from what is generally accepted as valid and established by the scientific community. According to the approach in this book, all superconductors are governed by the same principles. Electron–hole asymmetry is the key to superconductivity, without holes the Meissner effect cannot be explained and there can't be superconductivity, and BCS does not have the physical elements necessary to explain the Meissner effect.

Instead, according to what is generally accepted at the present time, there are the so-called 'conventional superconductors' that are perfectly well explained by BCS–electron–phonon theory, and a wide variety of 'unconventional' superconductors that are not yet understood and about which most scientists believe that they are described by the Hubbard model. The Meissner effect is understood by BCS since 60 years ago, and neither for conventional nor for unconventional superconductors are holes essential.

The second theme of this book is, assuming what I say in the first paragraph is valid, that bibliometrics completely fails in informing which are the most valuable scientific contributions in this field. The article by BCS, that says nothing about how the Meissner effect occurs, the most fundamental property of superconductors, has 7,956 citations; Koch's article, that identified for the first time 80 years ago the correct physical principle that explains the Meissner effect, has six. Slater's 1937 article, that identified for the first time the existence of mesoscopic orbits in superconductors, has only 14 citations (excluding mine), insignificant compared with the average 193 citations for each of the 127 articles published by Slater. Kikoin and Lasarew's article, that in 1932 identified for the first time the importance of holes in superconductivity, has a total of five citations excluding mine. Tokura, Takagi, and Uchida's article that announced that electrons rather than holes are the charge carriers in some cuprates, has 1,476 citations, even though it has been proven that this is not so. Anderson's article, predicting that the normal state of the cuprates is a 'spin liquid', has 6,002 citations even though it was proven shortly after its publication that the normal state of the cuprates is an antiferromagnet and not a spin liquid. The scientists of highest H index, with enormous number of citations to their papers on superconductivity, don't know the answers to the most basic questions in superconductivity, even more, they don't even know that those questions exist and are not answered. Conclusion: one has to take bibliometrics with many grains of salt: sometimes it helps, sometimes it doesn't help.

The third theme of this book is that scientific progress is often determined by accidents. If Kammerlingh Onnes in 1924 had used a solid rather than a hollow sphere in his experiment, he would have discovered the Meissner effect 9 years before Meissner did, when quantum mechanics had not yet been formulated, which potentially would have influenced its development. If Fröhlich had not hidden the true facts in 1950, perhaps the electron–phonon interaction would not have gained the universal prominence it did. Same if the isotope effect in ruthenium instead of in mercury would have been measured initially. If the compound MgB_2 would have been found in the 50's or 60's instead of in 2001, the importance of holes would perhaps have been recognized much earlier. If 30 K superconductivity had been found in the bismuthate BaKBiO, a compound with no traces of magnetism, before the first cuprate was discovered in 1986, instead of 2 years later, perhaps the scientific community would have focused on the crucial role of oxygen and not on the copper spin, to understand high temperature superconductivity. If Anderson had not written his RVB paper in 1987, a large number of brilliant young physicists would perhaps have dedicated their talents to real physics and not to fantasies during the last 30 years, and the field would have advanced much more. To keep in mind that the development of science depends on accidents like those, can help to keep the mind more open to the possibility that scientific truth may be different to what the scientific community believes it is at a given time.

Related to this, the fourth theme of this book is that sometimes established science can be wrong, and it is necessary to keep an open mind that this is possible and seriously consider alternative points of view. Not doing so protects the interests of the scientific establishment but delays the advancement of science to the detriment of humanity. There have been many historical examples of this, but always at a given present it is considered that that is a thing of the past, not of that moment, until after the fact.

Finally, the fifth theme of this book is that *simplicity* is often synonymous with truth in science. When explanations become complicated and only understandable to experts, like BCS, they tend to be wrong. Scientific truths are persuasive and easy to explain to non-experts, as I have tried to do in this book.

Until the scientific community is willing to acknowledge the clear anomalies [1, 2] confronting the conventional theory, the field of superconductivity will remain in its current state of suspended animation. And society will continue to be denied the benefits that would result from progress in this field. Then, some day in the future, all scientists will finally agree on what is the correct way to understand superconductivity in all materials. Until that time, I believe that readers of this book will understand *much more* about superconductivity than the so-called experts in the subject.

To review what was discussed in this book, a final exam for the reader: determine for each of the following items (closed book!) what is their relation to the title of this book, i.e. superconductivity either properly understood or misunderstood (or both), and explain in a few words why.

- *H*eisenberg
- *H*erbert

- *H*annes
- *H*ubbard
- *H*einz
- *H*eike
- *H*endrik
- *H*eroes
- *H*eretics
- *H*-field
- *H*oles
- *H*bar/2
- *H*elium
- *H*ydrogen
- H^-
- *H*all
- *H*g
- *H*Pd
- *H*-index

In conclusion, even if superconductivity properly understood does begin with holes, a lot remains to be understood. Assuming the non-conventional concepts in this book are valid, the correct microscopic theory will have to incorporate all those aspects. We don't have a complete microscopic theory that does this. Furthermore, it is necessary to design and perform experiments that confirm that the physics predicted by this theoretical approach is valid for all superconducting materials, as I have no doubt it is. And it is necessary to use those concepts to guide the search for new superconducting materials that superconduct at higher temperatures, even room temperature. Once this is achieved, superconductors will change the world, as semiconductors have.

I also believe that what superconductors are telling us about how quantum mechanics works at the macroscopic level will teach us new things about how quantum mechanics works at the microscopic level. Superconductors and the Bohr correspondence principle provide us with a powerful microscope that will allow us to understand quantum mechanics at the microscopic level in a new light. But that is a subject for another book, that somebody some day will write.

References

[1] A. Lightman and O. Gingerich, When do anomalies begin?, *Science* **255**, 690 (1992).
[2] J. E. Hirsch, BCS theory of superconductivity: The world's largest Madoff scheme?, arXiv:0901.4099 (2009).

Acknowledgements

I would like to conclude this book with some brief but deeply felt acknowledgements:

- To UCSD, for having provided me with the environment where I developed this work.
- To my wife Edith, for having accompanied and supported me and been patient with me during all these years.
- To my mother, rip, also a scientist, for having taught me to inquire and for her extraordinary interest in my research.
- To family, friends, and colleagues who supported me.
- To my collaborator from the beginning and during many years on this work, Frank Marsiglio, Professor at the University of Alberta.
- To Doug Scalapino, with whom I worked on superconductivity and other topics before starting this work and from whom I learned a great deal (as well as some that I had to unlearn).
- To the bulk of the scientific community, for not having paid attention to this work during all these years, thanks to which I could develop it without competitive pressures.
- To the few referees that have not objected and even recommended publication of these ideas in scientific journals over the years, usually not the highest profile journals, to some supportive journal editors, and to those that have invited me or recommended that I be invited to scientific conferences, seminars, and colloquia, where I could present these unorthodox ideas.

- To Chris B. Davis, for encouraging me to write a book on this topic, to World Scientific for agreeing to publish it, and to the World Scientific staff instrumental in publishing it.
- To the readers that devoted time and attention to read this book with an open mind.

Author Index

Subject Index

www.ingramcontent.com/pod-product-compliance
Lightning Source LLC
LaVergne TN
LVHW050928210926
848482LV00013B/571

* 9 7 8 9 8 1 9 8 3 7 8 8 5 *